Theory and Application of
MICROBIOLOGICAL ASSAY

Theory and Application of
MICROBIOLOGICAL ASSAY

WILLIAM HEWITT

Quality Control of Pharmaceuticals
Cheltenham, England

STEPHEN VINCENT

Pharmaceutical Microbiology
Glaxo Group Research
Greenford, England

ACADEMIC PRESS, INC.

Harcourt Brace Jovanovich, Publishers

San Diego   New York   Berkeley   Boston

London   Sydney   Tokyo   Toronto

ACADEMIC PRESS, INC.
San Diego, California 92101

United Kingdom Edition published by
ACADEMIC PRESS LIMITED
24-28 Oval Road, London NW1 7DX

50607

Library of Congress Cataloging-in-Publication Data

Hewitt, William, Date
   Theory and application of microbiological assay / by William
Hewitt and Stephen Vincent.
      p.   cm.
   Includes index.
   ISBN 0-12-346445-5 (alk. paper)
   1. Microbiological assay.   I. Vincent, Stephen.   II. Title.
QR69.M48H48   1988
576'.028—dc19   88-10424
                                                 CIP

PRINTED IN THE UNITED STATES OF AMERICA
88 89 90 91   9 8 7 6 5 4 3 2 1

CONTENTS

PREFACE

About half a century ago microbiological methods were introduced for the quantitative estimation of minute traces of certain biologically active substances. The recognition in the 1930s that some microorganisms were dependent on the presence of individual vitamins of the B group led to the development of assay methods for these growth-promoting substances. In the early 1940s the advent of penicillin on a commercial scale marked the beginning of the "antibiotic age" and the wide-scale use of microbiological methods for the quantitative estimation of growth-inhibiting substances.

Many papers have been published since the 1940s on assay methods for both growth-promoting and growth-inhibiting substances, with emphasis, generally, on problems with the individual substance to be assayed. Few books, however, have been published which present an overall view of microbiological assay in one source. The two volumes of *Analytical Microbiology: Theory and Practice* edited by Kavanagh (Volume 1, 1963 and Volume 2, 1972, Academic Press) provide a valuable reference to any microbiological analyst, giving a comprehensive treatment of theoretical aspects of agar diffusion and photometric assays as well as detailed guidance on a varied selection of assays. The volume *Microbiological Assay: An Introduction to Quantitative Principles and Evolution* by Hewitt (1977, Academic Press) gives guidance on general analytical principles, assay design, and calculation procedures.

This book was written to fill what we perceive as a gap in the literature, to serve as an introduction to newcomers to the field, and as a reference source for established workers in microbiological assay.

Inevitably, in writing the text we had to think about information that had previously been taken for granted. The result has been the inclusion of new material such as that on quality control of the assay itself and on the theory of assays for growth-promoting substances.

ACKNOWLEDGMENTS

We would like to express our appreciation to the many people whose help and encouragement contributed to making this work possible: the directors of Glaxo Group Research Limited, Greenford, Middlesex, UK for permission to utilize certain information and for authorizing practical assistance, including preparation of the art work; members of the Glaxo group—Mr. K. A. Lees, Dr. J. M. Padfield, Mr. J. P. Jefferies, and Dr. D. R. Rudd. Thanks are due to Mr. J. Drewe and Mr. E. Lacey of Autodata Limited, Hitchin, Herts, UK for supplying photographs for use in Figures 4.6, 4.13, and 4.14. Dr. G. D. Shockman of Temple University, Philadelphia, Pennsylvania kindly agreed to our use of his data to support our theoretical studies (Chapter 6). In Chapters 9 and 11 we have drawn freely on data published in the US Code of Federal Regulations (1985). Extracts from BS 1583, 1986 and BS 1792, 1982 (Chapter 3) are reproduced by kind permission of the British Standards Institution. (Complete copies may be obtained from the BSI at Linford Wood, Milton Keynes, MK14 6LE, UK.) Extracts from the British Pharmacopoeia (Chapter 3) are reproduced by kind permission of the Controller of Her Majesty's Stationery Office. Finally, thanks are due to Dr. F. W. Kavanagh, formerly of Eli Lilly and Company, Indianapolis, Indiana who first brought us together to collaborate in this venture and has continued to show great interest and make valuable suggestions.

CHAPTER 1

MICROBIOLOGICAL ASSAY: AN OVERVIEW

1.1 Introduction

Microbiological assay is a technique whereby the potency or concentration of a chemical substance may be determined by its effect on the growth of a microorganism. That effect may be to promote the growth of the microorganism by substances such as certain vitamins and amino acids, or to inhibit growth in the case of antibiotics and other substances having similar properties. The discovery that vitamins of the B group were essential for the growth of organisms of the *Lactobacillus* group and some other organisms led to the development of assay methods for these substances during the 1930s. The inhibitory effect of penicillin on staphylococci and other organisms had been known since 1929. However, it was with the early attempts to produce penicillin on a manufacturing scale in the early 1940s that microbiological assay of growth-inhibiting substances became established as an important analytical technique. The therapeutic value of penicillin was recognized before its chemistry had been elucidated fully and before any chemical method of assay had been developed. The same situation pertained in the case of many other antibiotics that were discovered and produced commercially, particularly during the following two decades. Thus, microbiological assay became established as an essential technique for process and product control in the pharmaceutical industry.

In the early days techniques were largely empirical, principles were not well understood, sources of error were unrecognized, and, as a consequence, assay results were often of poor quality. Despite the efforts and some successes of chemists in replacing microbiological assay with chemical or physical methods of assay, microbiological assay remains today the only practicable method of assay for many antibiotics and is a convenient method of choice in some cases where a chemical or physical method has been devised. Of course, as fundamental principles of the assay method have become better understood and sources of error have been recognized, the methods have become potentially more reliable. Evidence of the importance of microbiological assay in the quality control of antibiotics is borne out by the existence today of 40 or more international reference preparations for antibiotics.

Although chemical methods have been developed for vitamins of the B group, microbiological assay has in many cases the advantage of high sensitivity and/or specificity.

1

1.2   General Principles of Biological Assay

Microbiological assay has some features in common with biological assays in general, e.g., with *macro*biological assays, which involve the use of groups of living animals or of isolated organs. However, there are some important differences, which will be mentioned later. A fundamental principle of all true biological assays is that they depend on the comparison of the effect of a standard reference substance and a sample whose potency is to be determined on a "biological system." That biological system might be a group of mice or a culture of a microorganism. The comparison is between the *specific* biological activities of the two preparations, i.e., of the reference standard and of the sample of unknown potency. The same specific biological activity is contained only in preparations that are qualitatively the same. Thus, a comparison can be made between a tetracycline standard and a tetracycline sample, but not between a tetracycline standard and a streptomycin sample. The latter would not yield a meaningful potency estimate and in fact would give vastly different potency estimates according to the assay conditions.

In macrobiological assays such as those involving groups of animals, a very real problem that must be taken into consideration is the differing responses of individual subjects when exposed to the same stimulus. The economics of testing dictates that numbers of animals in a test shall be as small as is consistent with attaining an acceptable result. Despite efforts to standardize animals through breeding, random selection of animals to be allocated to groups receiving the same treatment, and perhaps crossover tests in which a group receiving a dose of standard on one occasion receives a dose of unknown sample on a subsequent occasion during a single assay, differences in mean responses to the same treatment remain. These differences result in a random variation of potency estimate which is known as "biological error." Pharmacopoeias describe assay designs intended to minimize biological error and provide calculation procedures to estimate the size of the error. These calculation procedures lead to an estimate of the width of fiducial limits (or confidence limits) for the estimated potency at a specified probability level, normally at $p = 0.95$. The statistical procedures also include some tests for invalidity of the assay, e.g., curvature of the response line that if found to be "statistically significant" should alert the analyst to possible invalidity of the assay.

Such statistical procedures that are described in pharmacopoeias are also intended for application to microbiological assays. Although there is certainly random variation in microbiological assay, its source is of a physical rather than biological origin. Consider a microbiological assay plate before incubation. It may have been inoculated with 10^8 microorganisms, which should be uniformly distributed throughout the agar. Any differences between individual organisms will be of no consequence. Thus the analytical microbiologist should look for sources of variation other than biological ones and endeavor to minimize them by appropriate techniques.

1.3   Basic Techniques of Microbiological Assay

In this work, attention will be focused on the two major practical techniques—the agar diffusion assay and the tube assay—which are applicable to both growth-inhibiting and growth-promoting substances. In the agar diffusion assay a nutrient agar gel is inoculated uniformly with a suitably sensitive test organism. Active substance is allowed to diffuse from an aqueous solution in a reservoir into the agar.

Upon incubation a zone is formed. The zone is one of inhibition in the case of growth-inhibiting substances and of exhibition in the case of growth-promoting substances. The width of the zone is dependent on the concentration of the active substance and provides the quantitative basis for the assay. Practical means of producing the zones are described in Chapter 4, and two different forms of zones are illustrated in Fig. 4.1.

In the tube assay the extent of growth of an organism in a liquid nutrient medium is dependent on the quantity of active substance that is added to and mixed uniformly with the liquid medium. The techniques for growth-promoting and growth-inhibiting substances differ in several respects which are described fully in Chapters 5 and 6. Growth of the organism is generally measured in terms of the turbidity produced although in the case of some vitamin assays the observed response is titration of the acid produced. The nature of the dose–response lines is quite different for the growth-inhibiting and growth-promoting substances. This, too, is described in Chapters 5 and 6.

Both these major techniques may be carried out manually and with quite simple equipment. Alternatively, many steps of both techniques can be automated or semiautomated. Automation of some steps such as the measuring and recording of zone size in the agar diffusion assay eliminates bias of human origin, which seems to be one of the common pitfalls in this method. It is not proposed to describe every available instrument or method, but to give basic guidelines in establishing sound microbiological and analytical principles together with examples of practical methods which have stood the test of time. Examples of equipment for automation are given in appropriate chapters.

1.4 Capability and Limitations

To minimize the effect of random variation of responses, replication of treatments is always necessary; e.g., typically six or more zones would be produced for each concentration of active substance in the agar diffusion assay. Similarly, three or more tubes would be used for each treatment in a tube assay. To achieve a high precision such as fiducial limits of $\pm 1\%$ ($p = 0.95$) would require a very much greater replication (and therefore more effort and cost) than to achieve a moderate precision such as fiducial limits of $\pm 4\%$ ($p = 0.95$). Of course, the degree of precision needed varies according to the purpose of an individual assay. There is a place for assays of high, medium and even low precision.

Microbiological assay (MBA) seems advantageous in that it measures the desired property, i.e., biological activity. For example, in the assay for a pharmaceutical preparation containing chloramphenicol that has partially decomposed, MBA will give a true measure of the undecomposed chloramphenicol. In contrast, a simple ultraviolet absorption assay would not distinguish between chloramphenicol and its decomposition products.

However, the measurement of biological activity may sometimes be misleading. Cyanocobalamin (vitamin B_{12}) may be assayed by the agar diffusion assay using *Escherichia coli*. This is an excellent method for the assay of pharmaceutical preparations formulated from pure cyanocobalamin. However, in the manufacture of vitamin B_{12} by fermentation, not only cyanocobalamin is produced but also several related substances having microbiological activity but not necessarily the required clinical activity. Other more specific MBA types may be applied to such mixtures such as the *Lactobacillus leichmannii* tube assay. When partially decomposed samples of tetracycline are assayed by the agar diffusion method using *Bacillus subtilis* there may be an apparent increase in potency. It is suggested by Garrett *et al.* (1971) that this is due to the production of anhydrotetracycline, which is less active therapeutically and more toxic than tetracycline. Its apparently higher potency in the agar diffusion assay may be due to greater diffusibility.

1.5 Current Status of Antibiotic Assays

The tendency for microbiological methods of antibiotic assay to be superseded by chemical or physical methods has been greater in Europe than in the United States. The differing viewpoints of two national authorities are contrasted in Table 1.1, which lists 36 antibiotics that are "official" in both Britain and the United States. The British Pharmacopoeia (BP,

Table 1.1

Pharmacopoeial Methods for Microbiological Assay

Antibiotic	Microbiological assay methods prescribed by:	
	British Pharmacopoeia (1980)	United States CFR (1980)
Amoxicillin	—	ad[a]
Amphotericin	ad	ad
Ampicillin	—	ad
Bacitracin	ad	ad
Candicidin	ad	t[b]
Capreomycin	ad	t
Carbenicillin	—	ad
Cephalexin	—	ad
Cephaloridine	—	ad
Cephalothin	—	ad
Chloramphenicol	—	ad, t
Chlortetracycline	ad, t	t
Clindamycin	—	ad
Cloxacillin	—	ad
Colistin	ad	ad
Colistin Sulfomethate	ad	ad
Cycloserine	—	ad, t
Demeclocycline	ad, t	t
Doxycycline	ad	t
Erythromycin	ad	ad
Gentamicin	ad	ad
Griseofulvin	—	ad
Kanamycin	ad	t
Lincomycin	—	ad
Neomycin	ad, t	ad
Nystatin	ad	ad
Oxytetracycline	ad, t	t
Penicillin	—	ad
Pheneticillin	—	ad
Phenoxymethylpenicillin	—	ad
Polymyxin B	ad	ad
Rifampicin	—	ad
Streptomycin	ad, t	ad,t
Tetracycline	ad, t	t
Tobramycin	ad, t	t
Vancomycin	ad	ad

[a] ad, Agar diffusion assay.
[b] t, Turbidimetric assay.

1980) prescribed MBA for only 20 of these, whereas the United States Code of Federal Regulations (CFR, 1980) requires it for all 36. The International Pharmacopoeia (1979) additionally describes MBA methods for dicloxacillin, novobiocin, and oxacillin which are not "official" in Britain or the United States.

1.6   When to Use Microbiological Assays

Whether for the assay of antibiotics or vitamins, whether by the agar diffusion or the tube method, a great deal of preliminary preparation is needed before even a single MBA can be carried out. However, once those preliminaries have been completed for one type of assay (e.g., a particular antibiotic by the agar diffusion method), it is a relatively simple matter to prepare for an additional antibiotic assay by the agar diffusion method. Microbiological assay, because it involves many steps, the majority of which must be completed in one day, is a technique best suited to teamwork. A team of three or more persons is potentially far more efficient than one person working alone. Thus MBA is an inefficient and extravagant method when very few samples have to be assayed. In contrast, a small team with an adequate work load can work efficiently. It follows that if the potential work load is low, then MBA is not the method of choice unless this is dictated by technical or pharmacopoeial considerations.

When facilities for MBA are well established there may still be a variety of points to consider before deciding between MBA, on the one hand, and chemical/physicochemical assay, on the other. It is not only a question of the nature of the sample to be examined but of the purpose for which it is to be examined. This will be made clearer by some examples that follow.

Example 1:

A manufacturer of bacitracin is selling bulk bacitracin to other pharmaceutical manufacturers for incorporation in pharmaceutical dosage forms. Microbiological assay is the only "official" method of assay and so this must be used with adequate replication to ensure a precise assay. A potency estimate that is appreciably higher than the true potency could lead to conflict with the regulatory authority. A potency estimate that is too low could result in financial losses when the selling price is related to potency and not to weight. It follows that the MBA must be carried out with sufficient replication to ensure a precise assay. (The relationship between replication and precision will be discussed in more detail in Chapters 8 and 12.)

Example 2:

A pharmaceutical processor purchases tetracycline hydrochloride for the manufacture of tetracycline capsules. Official assay methods for tetracycline and its pharmaceutical forms are MBA in both Britain and the United States (and probably so in most countries). The purchaser must assure himself of the quality of the bulk tetracycline. Thus, he must either be satisfied with the supplier's certificate of analysis issued by the supplier's own quality control laboratory, or a certificate issued by an independent analyst referring to a sample that has been properly taken and is representative of the containers/batches supplied, *or* he must have the material sampled and tested microbiologically in his own laboratory. In each of these possible cases, the material must be assayed with sufficiently high precision for the same reasons that were given in Example 1. Regarding the tetracycline capsules to be made from this bulk antibiotic, if the procedure consists merely of filling the capsules with the fresh bulk antibiotic alone, then no further assay is necessary. It is sufficient merely to ascertain that each capsule has been filled with the correct weight of antibiotic. This may be checked by following methods described in pharmacopoeias.

If, however, prior to filling the capsules the tetracycline hydrochloride were blended with some inert diluent to increase the bulk, then it would be desirable to check that the blending was satisfactory. Microbiological assay would not be an efficient and economically acceptable way of checking the blending. However, provided that the inert diluent did not contain ionic chloride, then a simple titration for chloride on the bulk blend or on filled capsules would establish whether the tetracycline hydrochloride was properly distributed.

Example 3:

A pharmaceutical processor purchases sodium benzylpenicillin and procaine benzylpenicillin for the manufacture of "fortified injection of benzylpenicillin." (This product generally includes in each dose: sodium benzylpenicillin 100,000 IU, and procaine benzylpenicillin 300,000 IU, together with buffer and excipients.) Again, the purchaser must ascertain by one means or other that the input antibiotics are of the required potency. Depending on local national regulations, for this purpose he may accept either chemical/physical or microbiological assays. However, to check that the ingredients are properly blended, regardless of national pharmacopoeias or other regulations, an assay for total penicillins whether by chemical/physicochemical or microbiological methods serves no useful purpose. Blending may be checked best by a determination of procaine base content of the mix, which is conveniently carried out colorimetrically. This might also be supplemented by a flame emission assay or atomic absorption assay for sodium, unless excipients include a substantial proportion of sodium.

If, however, the manufacturer wished to carry out long-term stability tests on the product or the laboratory of the regulatory authority wished to check total potency, then a total penicillin assay would serve a useful purpose. Either microbiological or chemical/physicochemical methods would be suitable. The regulatory laboratory should, in any case, ascertain that the proportions of procaine and sodium were correct.

Example 4:

A pharmaceutical processor manufacturing chloramphenicol suppositories must first ascertain that the input chloramphenicol is of the required quality. Chloramphenicol may be characterized adequately by chemical and physicochemical criteria, so that there seems to be no good reason to apply MBA to the input material unless so required by the law of the country. For control of the freshly manufactured dosage form, provided that there is no possibility of decomposition during processing, then the finished product may be checked for chloramphenicol content by means of its ultraviolet absorption characteristics. However, for stability testing a microbiological assay would be appropriate for the reasons given in Section 1.4. Precision of the assay should be sufficient to detect downward drift in potency during intervals of a few months. Again, for the reasons given in Section 1.4, assays by a regulatory laboratory on a sample taken from the market that has perhaps been subjected to conditions of storage that could result in deterioration, should be either a microbiological assay or a chemical/physicochemical method that distinguishes between chloramphenicol and its inactive decomposition products.

References

British Pharmacopoeia (1980). British Pharmacopoeia Commission, H.M. Stationery Office, London.
Garrett, E. R. *et al.* (nine authors) (1971). "Progression in Drug Research," Vol. 15, p. 271. Birkhauser Verlag, Basel and Stuttgart.
International Pharmacopoeia (1979). Vol. 1, 3rd ed. World Health Organization, Geneva.
United States Code of Federal Regulations (1980). The Office of the Federal Register, National Archives and Records Administration, Washington, D.C.

CHAPTER 2

TEST ORGANISMS

2.1 Introduction

Successful assays cannot be carried out without a proper understanding of the role and importance of pure, healthy, and robust test organisms in the form of bacterial or spore suspensions. Since microorganisms are one of the most important basic tools of the microbiological analyst, a certain amount of bacteriological know-how and practical experience in handling live cultures will be necessary. At least elementary knowledge of bacteriological theory and practice will be assumed from the reader. As far as possible full practical guidance will be given in easy-to-follow instructions to ensure satisfactory microbiological assays. It is not considered to be within the scope of this volume to provide full instructions in all aspects of bacteriological techniques such as media preparation, staining, and microscopy. For further aspects a good basic textbook of bacteriology should be consulted, such as Collins and Lyne (1984).

2.2 Safety

Commensal microorganisms are always around us in our daily environments and as a rule do not cause any disease. However, cultures grown in a laboratory can be present in very large numbers; e.g., a single bacterial colony may contain as many as a billion organisms. It should therefore be obvious that adequate precautions must be taken to ensure the safe handling of test organisms. An area often overlooked is the provision for safe disposal of finished or spent cultures and assay plates. Any accidental spillage must be immediately cleaned up using an effective disinfectant (see Appendix 1).

At the risk of stating the obvious, it is important to ensure that assay organisms are not carried out of the laboratory as contaminants on hands or clothing. Always wash your hands before leaving the assay laboratory. Use clean towels or, better still, disposable paper tissues. Protective overalls, properly fastened, should be worn during practical work. These overalls should be used only in the laboratory and they should be regularly laundered and kept in good repair.

Microbiological pipettes have cotton wool plugs, the main function of which is to protect the liquid inside the pipette from contamination. They do not provide absolute safety to the user. When diluting or pipetting bacterial suspensions, use a suitable teat or dispenser and do not pipette by mouth. Submerge the tip of the pipette just below the surface of the liquid, fill the pipette slowly, and expel the liquid gently to avoid the formation of aerosol. Surprisingly, dilute bacterial suspensions can be more dangerous than concentrated suspensions. This is because each droplet contains only a few organisms; after the moisture evaporates, the light particles remaining can stay in the atmosphere for long periods of time. Such small particles when inhaled can reach the lower respiratory tract and start an infection.

After diluting or transferring bacterial suspensions, pipettes must be placed in an autoclavable pipette jar, fully submerged in a freshly prepared effective disinfectant solution (see Appendix 1), and allowed to stand for at least half an hour before being washed in hot soapy water or water containing a suitable detergent (see Appendix 1). For extra safety, it is best to autoclave all soiled bacteriological pipettes, slides, etc., together with the pipette jars before washing up.

Further safety hints will be given throughout the text as and when needed to ensure the safe disposal of potentially infectious material and to protect all personnel.

2.3   Culture Media

Bacteria have a wide range of nutritional requirements for vigorous growth. It is not within the intended scope of this book to deal with this topic in depth. It will suffice to point out that some assay organisms, such as *Escherichia coli* or *Bacillus subtilis,* can be grown on media of very simple composition, while others, especially the lactobacilli used in vitamin assays, require much richer media. Successful growth of assay cultures should be achieved if the instructions given under the individual assays are faithfully followed. Commercially available dehydrated media make the life of the analytical microbiologist considerably easier. Several companies specialize in supplying both maintenance and assay media in complete formulation. Full instructions for preparation are provided on the label. These companies also supply free manuals which contain useful information such as details about their wide range of media and their uses. Suppliers of bacteriological media are listed in Appendix 2.

2.4 Freeze-Dried Assay Cultures

When setting up a microbiological assay laboratory, it will be necessary to obtain assay organisms. These are available from the various national culture collections, as indicated under the individual assays. New cultures will invariably arrive freeze-dried in a glass ampul, which may be under vacuum. Instructions for opening and culturing are usually supplied with the ampul, which should be followed. Some ampuls have a separate small cotton wool-plugged tube inside. Instructions for this type are given here, as an example. Score the ampul all around with an ampul file above the label and apply a red-hot glass rod or electrically heated hot wire to the file mark until the glass cracks. If it does not crack the first time, it will be necessary to repeat the application of the reheated red-hot glass rod until a distinctive cracking sound is heard. Allow time for air to enter the ampul; then, carefully remove the top of the ampul and discard it into disinfectant solution. Carefully tip out the small inner tube containing the culture and with the aid of forceps loosen the cotton wool plug. Discard the now-empty bottom half of the outer ampul into the disinfectant solution.

Fit a small rubber teat to a sterile Pasteur pipette and withdraw a small volume (~ 0.5 ml) of recovery broth into the pipette. Tryptone soy broth is used most frequently for this purpose, but other media may have to be used for the more fastidious organisms (see Appendix 3 for formulas and preparation), for which full details will be given under the appropriate assays. Withdraw the cotton wool plug from the small tube and slowly introduce a few drops of broth into the tube. Make sure you do not overfill the tube. Reconstitute the culture by repeatedly filling and ejecting the broth with the pipette. Three or four applications are generally sufficient to resuspend the culture. Two or three drops of the resuspended culture are then spread on the surface of solid agar (e.g., nutrient agar) in a petri dish and the remainder is added to 10–20 ml of broth. Discard the cotton wool plug, the small tube, and the Pasteur pipette into disinfectant solution, and place the cultures in an incubator. Reconstituted cultures should be incubated for at least twice the usual growth period of the organism before pronouncing them nonviable. Every culture must be clearly identified with the species name and number and the date of inoculation. As soon as turbidity is evident in the broth after incubation, subculture a small volume with a sterile Pasteur pipette into fresh sterile medium and reincubate. Some organisms will need at least three successive subculturings before they regain their full physiological characteristics after freeze-drying. The inoculated agar plate should show distinct uniform colonial forms.

Cracknell (1984), curator of a large culture collection, reports that a number of people, especially those with limited bacteriological experience, often fail to recover freeze-dried organisms in broth and claim that the culture supplied was nonviable. It is therefore useful to employ in addition solid medium, on which the success rate of recovery is generally much higher.

If satisfactory growth was obtained in the broth, then the agar culture can serve as a check on the purity of the strain, and a single colony could be selected for culture maintenance (see Section 2.6).

If, on the other hand, the broth culture failed to grow up for some reason, the agar plate should provide a check for both viability and purity.

2.5   Culture Checks

Having successfully recovered the assay organism, the next step will be to check the purity and the identity of the culture. Just because growth was obtained in the broth, it does not prove that the desired assay organism was recovered. In inexperienced hands broth cultures can be very easily contaminated, and it is possible to finish up with a mixture of organisms; sometimes, an adventitious organism can actually take over completely. Therefore, it is now necessary to streak out a loopful of the broth culture on the surface of a suitable solid medium in a petri dish. After incubation for 16–48 hours at the optimum temperature for the organism, examine the developed colonies. With practice, it will be possible to achieve separation of the culture into distinct colonies without touching or confluent growth. The developed colonies should look uniform in appearance with regard to shape, color, size, etc. Slight variation in colony size may be permitted, but if more than one distinct colony form is discernible the culture probably became contaminated and it will be best to start afresh with another freeze-dried ampul.

Prepare a Gram stain (see Appendix 7) of a typical colony and examine it under the microscope using an oil immersion objective with a magnification of 100:1. Those interested in systematic bacteriology should consult "Bergey's Manual" (Krieg *et al.,* 1984; Sneath *et al.,* 1986) and Cowan and Steel (1970). Use may be made also of bacterial identification kits such as the API system of identification (API Laboratory Products Limited, Grafton Way, Basingstoke, Hampshire RG22 6HY, England), where applicable. The various API kits consist of a strip of plastic microtubes containing dehydrated biological reagents. The freshly grown pure culture of the organism to be identified is suspended in the inoculation medium supplied with the kit, and this inoculum is distributed into the tubes using a sterile Pasteur pipette. After incubation and addition of a few further reagents, the positive and negative reactions are recorded and (using the octal coding system) a seven-digit profile number is obtained. Identification is then usually possible either by means of the Analytical Profile Index obtainable from API, or a differential chart listing the various biochemical reactions or by computer through the offices of the manufacturer. Identification is usually possible within 24–48 hours using the kits.

2.6   Culture Maintenance

The next step will be to establish a system of culture maintenance, including master cultures, submasters, and working cultures. Master cultures only carry the strain. Their job is to preserve the culture with all its original characters, and only in an emergency are they called on to provide another function, such as rescuing the strain in case of contamination or loss of strain. Submasters provide the stock from which weekly or daily cultures are set up as needed. It is a sound practice to keep all these cultures on sloped agar media, called "slopes" or "slants," in universal bottles or test tubes, provided with screw caps. Blood agar base (see Appendix 3) is satisfactory for most organisms, but some will require specialized media. For example, some media may have to contain special growth factors, others specified amounts of an antibiotic so as to ensure continued resistance of the assay culture to the particular antibiotic. Appropriate media will be recommended under the individual assay details for each test organism.

It is common practice to subculture master and submaster cultures at monthly intervals, but there are, of course, always exceptions; e.g., yeasts may not need to be subcultured more often than once or twice a year, while some of the more sensitive bacteria will need weekly subculturing. Figure 2.1 shows the typical steps involved in setting up a culture collection.

The lactobacilli used in the vitamin assays are much more exacting in their growth requirements, and for best results it is recommended to passage them alternately on solid and liquid media, i.e., microinoculum agar (Difco, see Appendix 3) one week, followed by microinoculum broth (Difco, see Appendix 3) the next, then back to the agar the third week, and so on. For best results the following procedure is recommended:

(1)   Grow up the freeze-dried culture after reconstitution in microinoculum broth, and subculture into fresh microinoculum broth on at least 3 successive days.

(2)   With a sterile Pasteur pipette, transfer sufficient freshly grown culture from the microinoculum broth into the top of a microinoculum "deep" (or "stab") to cover the surface. Heat a bacteriological wire loop (or straight wire) in the flame to red glow and after cooling, stab the agar

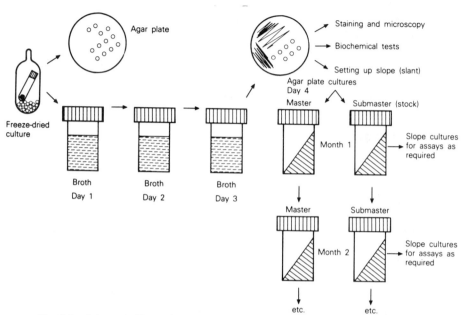

Fig. 2.1. Schematic illustration showing the usual steps involved in setting up a typical culture maintenance program starting with a freeze-dried culture. The time interval required between subculturing operations is dependent on the growth characteristics of the test organism and may vary from 1 week to several months.

repeatedly, so that some of the broth is carried below the surface of the agar.

(3) Incubate at the appropriate temperature for 16–18 hours and if there is visible growth, transfer to the refrigerator at 4°C.

(4) The following week, scoop out some of the growth from the depth of the agar with a sterilized bacteriological loop, and transfer this growth into a fresh sterile microinoculum broth. After 16–18 hours of incubation, visible turbidity should be observed in the broth. Keep this culture in the refrigerator at 4°C until the next week, then return to step 2 above; keep to this routine of alternating between solid and liquid media. This method is very satisfactory for maintaining a vigorously growing culture, but the utmost care is needed to follow aseptic techniques so as to avoid adventitious contamination.

2.7 Preparation of the Inoculum

For assays requiring nonsporing organisms, such as *Staphylococcus aureus, E. coli,* and *Micrococcus luteus,* set up a subculture from the submaster on an agar slope the night before the assay and incubate for 16 – 18 hours. The following morning, wash off the growth in a convenient amount of sterile saline or 0.1% peptone diluent (see Appendix 3) to obtain a heavy, milky looking suspension. If all the growth is not removed easily by gentle rocking or shaking, sterilize a bacteriological wire loop and loosen the growth carefully with the loop, taking care not to chop up the agar slope.

Safety note: Remember to sterilize the loop after use.

The suspension can be kept in a sterile bottle closed with a screw cap for at least 1 week (in some cases as long as 6 – 8 weeks), provided it is stored in a refrigerator at 4°C.

It will now be possible to standardize the suspension for the agar diffusion assay. Add aliquots of 0.5, 1.0, 1.5, and 2.0 ml to standard volumes of molten assay agar which has been cooled to about 48°C and, after blending without the formation of excessive air bubbles (best achieved by a rolling and tumbling action of the bottle — but not shaking), pour the inoculated medium immediately into previously leveled plates. Burst any air bubbles quickly by flicking the Bunsen flame across the surface before the agar sets. After allowing the agar to solidify and after about 2 hours of storage at 4°C (see Chapter 4 for more details about plate preparation), plate out two or three dose-level standard solutions in triplicate and incubate all the plates at the required temperature overnight.

The next morning, examine the developed zones of inhibition or exhibition for "definition" (i.e., clarity of zones, sharpness of zone edges, background growth), and measure the zone diameters carefully. For an acceptable assay, one should have clearly defined zones, with zone edges preferably highlighted by a thin layer of extra heavy growth, and in the case of antiinfective assays, the clear, sharp zones should be surrounded by a smooth, opaque, and continuous growth. Vitamin assays should have sharply defined zones of growth or "zones of exhibition" with no or only minimal background growth. In two-dose level assays ideal zone diameters using 8-mm-diameter agar wells should be around 18 mm for the low and about 22 mm for the high level. This is only a rough guide, but zones in excess of 28 mm may get "pulled" or distorted and may even touch or fuse. Zones less than 14 mm will be difficult to measure accurately, and potency estimates will become uncertain. In the majority of cases a 2:1 dose ratio will be possible, but on a few occasions a 4:1 dose ratio may be necessary. It might be a good idea to plot the average zone diameter against the logarithm of concentration used. This should give a clearer indication about the choice of future levels to be selected. The inoculum levels giving the above-described best conditions can be used in subsequent assays for the next 4 or 5 weeks, provided the suspension is kept at 4°C in a refrigerator.

If the zones are very small and the background growth is heavy, it may be necessary to dilute the original suspension 1:2, 1:5, 1:10, and 1:20, and set up test plates again with varying amounts of inocula from each dilution before the most satisfactory level of inoculum is found. Make a note of all the necessary steps to achieve optimum assay conditions and standardize your procedure for all future assays. It is surprising how easy it is to establish standard operating procedures without the use of expensive instrumentation.

Naturally, it is permissible to use more sophisticated methods, such as opacity tubes for visually standardizing bacterial suspensions, or a nephelometer or spekker to obtain accurate optical measurements, but establishing and following simple practical but standardized operating procedures are more likely to lead to success than heavy expenditure in gadgets. It is far more important to ensure that the assay organism was grown up freshly from a viable culture on the correct medium and for the correct length of time, than to finish up with an accurate optical measurement on a nonviable or contaminated suspension.

The aim of this book is to introduce simplified, practical, and in most cases inexpensive procedures that have come from many years of experience and have stood the test of time. It is very important to grasp the essentials right at the beginning and to understand clearly the mechanics of the assay (any assay); once one assay has been shown to work satisfactorily, other assays can be added to the repertoire relatively easily, since the same basic tools and techniques can be utilized again and again with slight variations. An assay laboratory can be set up for a moderate expenditure and once the basic equipment has been acquired, the day-to-day running costs will be relatively low.

Preparation of the inoculum for vitamin assays is very critical, and full instructions will be provided under each individual assay. The methods will initially involve vitamin enrichment steps to ensure rapid and dense growth of the assay organism. These are followed by depletion steps to minimize carry-over of the vitamin to be assayed into the assay medium and to eliminate the need for centrifugation.

2.8 Preparation of Spore Suspensions

The most commonly used assay organisms are *Bacillus subtilis, B. pumilus,* and *B. cereus,* all of which are aerobic spore-bearing bacilli. Instead of using these organisms in their vegetative phase, it is more convenient to allow them to sporulate under optimum conditions. The main advantages of using spore suspensions for assays are listed briefly:

(1)   They are relatively easy to produce.

(2)   Once prepared, they can be standardized; if kept in the refrigerator, set quantities per standard volume of assay agar can be used for long periods — usually months and with occasional slight adjustment for years — practically guaranteeing successful assays every time.

(3)   The use of spore suspensions eliminates culture maintenance and regular subculturing altogether.

The various published media and methods of producing spore suspensions are legion and for a beginner they could be more confusing than helpful. The method proposed here has the merit of being simple to follow; also, it works for all strains of bacilli and produces a stable spore suspension. The spores are very easy to harvest, requiring neither centrifugation for initial concentration nor repeated washings followed by further centrifugation. The actual concentration of spores may not be as heavy as on some more complex media; on the other hand, ease of preparation and handling more than compensate for this slight disadvantage, especially in that a fresh spore suspension can be readily made available in about 8 – 10 days from starting.

The formula and the preparation of the sporulating medium is given in Appendix 3. It is important to ensure that the medium is allowed to solidify after sterilization in such a way that the largest possible surface area is produced, because copious spore production requires a plentiful supply of oxygen from the air.

For successful preparation of spore suspensions follow the simple steps given here:

(1)   Inoculate about 20 ml of soybean-casein digest medium (see Appendix 3) at the start of the working day, preferably by "picking off" a single typical colony grown on a blood agar base (see Appendix 3) plate overnight. (Later on it will be possible to inoculate the broth from a previous spore suspension directly, provided it is known to be pure and free from contamination.)

(2)   After about 6 hours of incubation at $31° - 35°C$, shake to suspend the culture and transfer 1 ml onto the surface of each bottle of the sporulating medium. (For example, one can use 200-ml medical flat bottles with screw caps.) Tilt the bottles slowly backward and forward and from side to side to make sure that the whole surface is covered by the broth, then transfer the inoculated bottles to the incubator. Place the bottles on their sides with the inoculated agar surface at the bottom and parallel to the shelf. Loosen the screw caps about half a turn, so that air can circulate freely without exposing the medium to aerial contamination.

(3)   After about 8–10 days of incubation, 90% of the culture should have turned into spores. It is advisable to check the proportion of spores in a few of the bottles by carrying out a spore stain (see Appendix 7), followed by microscopic examination. When 90–95% of the culture consists of spores, they are ready to be harvested.

(4)   Aseptically introduce about 10–15 ml of sterile saline (see Appendix 3) into the first bottle, and gently run the liquid over the surface back and forth and up and down. The surface growth should simply fall off the agar when the bottle is stood upright. Transfer this opaque liquid into the second bottle and wash off the growth as before. Collect the washings in this manner from as many bottles as possible, that is, until the suspension becomes too thick to manage. Start with another bottle using fresh saline, and again collect the surface growth from at least two to three bottles. Pool the collected spore suspensions together into a sterile bottle. The pooled suspension should be a heavy, thick liquid with a muddy, murky appearance.

(5)   Pasteurize the suspension at 80°C for 20 minutes and distribute it in 20-ml amounts into sterile universal bottles with screw caps. Set up test plates by inoculating 0.5, 1.0, and 2.0 ml of the suspension per 100 ml of the appropriate assay medium, then continue as for the vegetative bacterial suspensions, by plating out the appropriate standard solutions. Depending on the number of spores present, the formulation and thickness of the assay medium, the temperature of incubation, and the actual antibiotic solutions used, clear zones of inhibition should be evident after overnight incubation at one of the spore concentrations used. It may be necessary to dilute the spore suspension further, say 1:5, 1:10, 1:20, or even 1:100, with sterile saline before optimal conditions are found. Once the best quantities and dilutions have been ascertained, label all the small bottles of the spore suspension with the relevant details, including name of the organism, dilution and amount to be used per 100 ml, and date. Permanent records should also be kept of each assay suspension, results of tests, definition, etc.

(6)   If the zones are too large, touching or diffuse, and the background growth is too sparse, then the spore suspension is too weak. Perhaps the growth was not profuse enough, or the rate of sporulation was insufficient or, maybe, too much saline was used for harvesting the spores. It may be possible to salvage and concentrate the spores by centrifugation and resuspension in a smaller volume of diluent, but generally speaking, it would be more sensible to start afresh with a new culture.

(7)   If there are very small or no zones at all, and the background growth is dense, then the spore suspension requires further dilution as indicated under step 5. In extreme cases, a contaminant may have taken over which is

resistant to the antibiotic being tried, although this is less likely if the suspension was pasteurized as described under step 5 above.

(8)   If the contaminant was a nonsporing vegetative organism, it would have been killed by the pasteurization treatment, and consequently there would not be any growth on the assay plate, nor would there be any zones. The obvious course of action would be in this case to start again with a fresh culture.

2.9   Preliminary Assay Considerations

Now that the culture suspensions for the various assays have been prepared, it is time to consider how to put them to work. From a practical point of view it is best to regard them as important reagents that had to be carefully standardized so that they could be of help in the final and most important step, i.e., in the accurate measurement of the developed zones.

However, let us first look at the difference between the organisms used for the assay of antiinfectives and those used for the assay of growth-promoting substances. In fact, the difference can be stated in simple terms: the organisms used for assaying antibiotics and preservatives are generally speaking nonspecific, whereas the ones for the assays of vitamins and amino acids tend to be specific. This statement can be put in another way by saying that several different microorganisms can be utilized for assaying the same antibiotic, but the assay of growth-promoting substances is usually restricted to those microorganisms that require the particular vitamin or amino acid under test as a growth factor. A number of these species were found to require growth factors for their development in their natural habitat; others were induced by artificial means to become dependent on the necessary growth factor.

2.10   Minimum Inhibitory Concentration

One of the simplest methods for gaining some idea of the effectiveness of an antibiotic's action is to determine what is known as minimum inhibitory concentration (MIC). There are two main methods depending on the substrate used: broth and agar.

2.10.1   Broth Dilution Method

A convenient starting concentration of the antibiotic is made up in nutrient broth or any other suitable liquid medium, and then twofold dilutions in broth are prepared serially. The most frequently covered range is from 0 to 100 μg/ml. Inoculate each tube with one drop of the bacterial suspension of about 10^6 colony-forming units (CFU). After incubation the MIC level is assessed visually. The first clear tube after turbidity, starting with the blank broth, is recorded as the MIC. In other words, the highest dilution of the antibiotic preventing growth is taken as the MIC of the test organism.

It must be stressed that the MIC varies from organism to organism and from strain to strain of the same species; it may give different values in different media or, indeed, from day to day. The end point of the MIC is also greatly affected by the density of the inoculum and by the precision of the dilution technique, as small errors committed at each step are multiplied out. In the example given the error would double at each step. For the same reason, a change of MIC by only a single tube in the series would result in either halving or doubling the previous value. Clearly, this technique is not suitable for the potency determination of unknown samples.

The most important application of this method is found in hospital laboratories, where the MIC values of clinical isolates give important guidance to physicians in the antibiotic therapy of patients.

2.10.2 Agar Dilution Method

In this method the serial dilution of the antibiotic is prepared in molten nutrient agar or other specialized agar medium depending on the nature of the test cultures, and then plates are poured. The surface of the plates is dried at 37°C for not more than 1 hour and inoculated with about 20 μl of an overnight broth culture containing 10^5 CFU. Several organisms can be tested on the same plate (i.e., at the same antibiotic concentration), and a large number of cultures can be screened rapidly, especially if a multipoint inoculator is used.

After incubation at 37°C for 18 hours, the level with no growth or with the development of only a few discrete colonies is taken to be the MIC of the particular strain under test. Agar plates not containing any antibiotic must be included as a growth control, and a control strain of known sensitivity should also be present on each plate with the other organisms.

The main application, again, is for the checking of infectious clinical contaminants to determine which antibiotic is most suitable for treatment.

It will be shown in later chapters that only methods based on comparison with a standard material can be considered as satisfactory for the estimation of samples with unknown potency. The turbidimetric determination of both antibiotics and vitamins can be regarded as an extension of the tube dilution method, though employing a very limited range with closely spaced concentrations of active substance.

References

Collins, C. H., and Lyne, P. M. (1984). "Microbiological Methods," 5th ed. Butterworths, London.
Cowan, S. T., and Steel, K. J. (1970). "Manual for the Identification of Medical Bacteria." University Press, Cambridge, England.
Cracknell, P. M. (1984). Personal communication.
Krieg, N. R., *et al.,* eds. (1984). "Bergey's Manual of Systematic Bacteriology," Vol. 1. Williams & Wilkins, Baltimore/London.
Sneath, P. H. A., *et al.,* eds. (1986). "Bergey's Manual of Systematic Bacteriology," Vol. 2. Williams & Wilkins, Baltimore/London, Los Angeles, Sydney.

CHAPTER 3

TEST SOLUTIONS

3.1 Introduction

The basis of all microbiological assays, whether of growth-inhibiting or growth-promoting substances, is the comparison of the effect of a sample of unknown potency (the "unknown") with a reference standard of defined potency on the growth of a suitable test organism. Test solutions are the final dilutions at a series of two or more concentrations that have been prepared from reference standard and unknown for application to the test system — the assay plate or tube. Ideally, test solutions prepared from reference standard and unknown should be qualitatively identical and differ only quantitatively. This is because the comparison that is being made is the quantitative difference in the *specific activities* of the two preparations, i.e., of the activity that is specific to the reference material. In practice, this means that it is first necessary to select the right chemical form of the reference standard, to treat the unknown in such a way that potentially interfering substances are removed or their influence neutralized, and to carry out dilutions with such accuracy as to ensure that differences in the specific activities of the starting materials (reference and unknown) are faithfully reproduced in the various dose levels of the test solutions.

3.2 Reference Standards: Principles and Problems

As microbiological assays are based on a comparison of the effects of an unknown and a reference standard on a test system, it is evident that if assay results are to be universally comparable, ideally, all reference standards for a single active substance should be identical. Because it is not practicable for all laboratories throughout the world to use a single standard for each substance in everyday assays, it is generally feasible to aim for standards that are qualitatively very similar and whose potencies can be defined in terms of a universally accepted single reference material for each substance.

There is a general tendency for scientists to aim to replace biological methods (based on a comparison with a reference standard) by methods based on determination of chemical or physicochemical characteristics. These may be absolute methods, dependent on the known chemical structure of the active substance, or its known physical characteristics such as ultraviolet absorption. Alternatively, chemical and physicochemical characteristics may be compared with those of an authentic specimen of the substance. Sources of such authentic specimens are listed in Appendix 5.

Even in those cases where chemical or physicochemical methods have been devised, the comparative biological methods survive; they are found convenient and continue to receive official recognition in many countries. A striking example is in the case of the natural and semisynthetic penicillins for which, in Europe, chemical and physicochemical methods are now "official." By way of contrast, in the United States the Code of Federal Regulations (CFR) still recognizes microbiological methods of assay.

It is a fundamental principle of biological assays that reference standard and unknown be qualitatively similar (or yield the same molecular or ionic species in the final test solution). The reader is referred to the principles expounded by Miles (1952), Lightbown (1961), and Wright (1971), and summarized by Hewitt (1977).

For those antibiotics for which chemical methods of analysis either have not been devised or have not yet gained official recognition, microbiological assay remains the only acceptable test method and so assays must be based on a comparison with an appropriate reference material. Quoting from the British Pharmacopoeia (BP 1980, p. A121), reproduced here by kind permission of the controller of Her Majesty's Stationery Office:

> Standard Preparations . . . are of two kinds, primary standards which are established, held and distributed by the appropriate international or national organization and secondary (working) standards which are preparations the potencies of which have been determined by an adequate number of comparative tests in relation to the relevant primary standard.
>
> A primary standard is a selected representative sample of the substance for which it is to serve as a basis of measurement. It is essential that primary standards shall be of uniform quality and as stable as possible; these conditions are usually ensured by providing the preparations in a dry state, dispensing them in sealed containers free from moisture and oxygen, and storing them continuously at a low temperature and in the absence of light.

The ultimate reference standards for many substances that are standardized biologically, such as antibiotics, are the International Biological Standards and International Biological Reference Preparations. To generalize, the first of these two groups are those substances to which an international unit (IU) has been assigned and the second refers to those substances for which no international unit has been assigned. In the interest of brevity, both these groups will be referred to here henceforth as international biological standards. These international standards for antibiotics may be obtained from the International Laboratory for Biological Standards, National Institute for Biological Standardization and Control, Blanche Lane, South Mimms, Potters Bar, Hertfordshire EN6 3QG, England, where they are prepared and held for the World Health Organization. However, they are available only in small quantities and are provided free of charge to the laboratories of national regulatory authorities. There are currently about 40 international biological standards for antibiotics.

A quite substantial effort is needed to calibrate even one subsidiary standard with the required precision and so, with some notable exceptions, few countries have made much progress in establishing national standards. Britain has its national reference standards for antibiotics, most of which are identical with the corresponding international biological standards, having been prepared in the same laboratory from the same batch of material. The United States has an extensive set of United States Pharmacopoeial (USP) Reference Standards for medicinal substances in general, which includes antibiotics and vitamins. These may be purchased by laboratories of all countries.

A start has been made in some of the more advanced of the developing countries in the establishment of national standards for antibiotics by calibration of material against the corresponding international biological standards. Turkey and Thailand are examples of such countries.

There is now a growing awareness that collaboration between nations to set up regional reference standards is economically more viable. In Europe, a start has been made on European Pharmacopoeia Standards. In Asia, there is collaboration under the auspices of South East Asia and Western Pacific Regions of the World Health Organization toward the establishment of regional standards. In WHO's Eastern Mediterranean Region, assistance is being given for the development of a National Reference Standards Laboratory in Egypt, which it is envisaged will eventually fulfill a regional role. Initial plans are that a sufficient number of vials of each standard be set aside to satisfy the needs of all Egyptian users for a period of several years.

Major pharmaceutical manufacturers, having the need for reference standards in large quantities and having the technical and economic capability, set up their own in-house standards. These might be calibrated against the international or national reference standard with a precision of $\pm 1\%$ ($p = 0.95$).

In the absence of a recognized national or regional standard, the individual laboratory is faced with the problem of setting up its own working standards. This is a very widespread problem, the solution to which must be dependent on the function of the laboratory (e.g., laboratory of a regulatory authority or a manufacturer's quality assurance laboratory) and on local circumstances.

If the laboratory of the regulatory authority finds a sample from the market to be substandard, its finding might be questioned by the vendor or manufacturer. In that case, unless the reference standard used by the regulatory laboratory in the assay has itself been properly and convincingly calibrated against the corresponding international standard, then the regulatory authority would find it difficult to take any formal action to recall the offending batch that the sample represents.

However, even when such a proper working standard is not available, the regulatory laboratory might still perform a useful function by informal comment on the quality of the sample. Certainly many manufacturers will respond to informal comments. By way of contrast, the reports of a quality control laboratory of a pharmaceutical manufacturer are, in general, solely the internal concern of the company itself. Any working standard will suffice provided that it enables the quality control laboratory to ensure that the company's products are found to be of the required quality if examined independently by the government regulatory laboratory.

Reference standards for microbiological assay may be divided into two groups:

(1)   Those substances that can be calibrated by comparison with a chemical reference substance or through their chemical and physicochemical characteristics alone, by means of chemical or physicochemical methods of analysis.

(2)   Those substances that can only be calibrated by means of a microbiological assay relating their potency to that of the international reference standard.

The first of these two classes presents problems that are relatively easy to resolve. International Chemical Reference Substances, if needed, are somewhat more freely available than are the international biological reference standards. The chemical and physicochemical methods for characterization and calibration make it possible to assign to a working standard a potency or percentage strength or purity with adequate precision.

When the substance can only be calibrated by means of microbiological assay, then the problems are considerably greater. Let us suppose that a working standard should be calibrated against the international standard with an error no greater than $\pm 1\%$ ($p = 0.95$). If a single assay leads to confidence limits of $\pm 5\%$ ($p = 0.95$), then we can say that it will require roughly 25 such assays to yield a mean potency estimate with confidence limits of $\pm 1\%$ ($p = 0.95$). As a crude guide, precision is inversely proportional to the square of the replication.

However, it is preferable that the 25 or so assays be carried out in more than one laboratory so as to bring to light any bias that might arise in the estimates from an individual laboratory. The organization of a collaborative assay itself produces problems, which, although they can be resolved, takes time.

3.3   Reference Standards: Practical Approaches

Since the establishment of a working reference standard—especially one that must be calibrated biologically—necessitates a substantial effort, it is prudent to select material that when properly packaged and stored will retain its potency without readily detectable loss for 3 or more years. In selecting suitable material, the following factors should be taken into consideration:

(1) *Purity and qualitative similarity with the ultimate standard.* However, a different salt that will yield the same active ionic species when prepared as a test solution may be acceptable.

(2) *Moisture content.* An anhydrous substance is generally more stable. For primary standards, it has already been stated in Section 3.2 that anhydrous materials should be used (a B.P. 1980 recommendation). However, an anhydrous substance may be hygroscopic enough to make moisture pickup after opening the package and during weighing a significant factor. For this reason there is a growing tendency to assign potencies for working standards on the "as-is" basis, i.e., on the basis of the material containing a small percentage of moisture. The rationale behind this tendency is that upon storing the material in a refrigerator, loss in potency due to the presence of 2–3% moisture will be low and will present less of a hazard than possible moisture pickup by anhydrous materials. However, it would be unwise to generalize for all substances. Each case should be considered on its own merits.

Some suggested chemical forms suitable for working standards are presented in Table 3.1.

For any working standard, an appropriate quantity of the selected material should be obtained from a reliable source such as a major manufacturer of the substance. The quantity might be, for example, from about 20 g to several hundred grams, according to the number of laboratories to be served and the expected frequency of use of the working standard.

It is naturally necessary to establish first the identity and purity of the material through application of the pharmacopoeial tests, which include thin-layer chromatography, spectrophotometry, and other physicochemical methods. Even though the material may be calibrated by means of a microbiological assay, it must be ascertained that it does not contain undue quantities of impurities that might interfere with its specific biological activity. The manufacturer's own analytical data might be offered. Additionally, the material should be reexamined as fully as the recipient laboratory's facilities permit. If chemical and physicochemical testing indicate that the material is suitable for use as a working standard, then a start may be made on the more laborious and therefore more expensive task of microbiological calibration. Ideally, the working standard should be calibrated so that a potency may be assigned with confidence limits of $\pm 1\%$ ($p = 0.95$). However, the time and cost involved may be such that some compromise should be made.

Suppose that about 100 g of material were obtained and had been shown to be suitable for use as a working standard and that a potency had been assigned. Approximately 1-g quantities could be placed in 20 vials and the remainder stored in a screw-capped, wide-mouthed powder bottle. It is not necessary to weigh accurately the quantities put into the vials. A standard measuring scoop may be devised so that the transfer of material can be completed rapidly with minimal pickup of moisture. Vials and the larger container should be well sealed, and then labeled with the name of the substance and a reference number. They should then be placed in an outer container such as a desiccator or a polyethylene box with a well-fitting lid and containing self-indicating silica gel. The containers should be kept in a refrigerator or cold room. In this way, the bulk of the material (the 80-g portion) is stored compactly and its container is opened only infrequently. After opening, a vial of the working standard may be used on several occasions provided that exposure to air and moisture is kept to a minimum. A logbook should be kept giving the following data:

Reference number allocated by the laboratory.
Exact description of the substance.
Date obtained.
Origin: the supplier and manufacturer.
Quantity received and description of the package in which received.
Date of manufacture.
Manufacturer's analytical report, if available and date.
Laboratory's own analytical data (full tests) and potency assigned, preferably with confidence limits. Date of tests.

3.4 Preparation of Test Solution

Test solutions have already been defined in Section 3.1 as the final dilutions at a series of two or more concentrations that have been prepared from reference standard and unknown for application to the test system. The very high dilutions that are necessary in the majority of assays is a factor that poses some special problems in attaining accurate and reliable potency estimates. To illustrate the extent of dilution that may be required, some extreme low-dose concentrations are quoted.

Table 3.1

Substances That May Be Used as Working Standards for Antibiotic Assay

Chemical form	Typical potency[a]	Suitable for the assay of:	*Not* suitable for the assay of:
Amoxicillin trihydrate	885 μg/mg	Amoxicillin trihydrate	
Amphotericin	940 IU/mg	Amphotericin	
Ampicillin trihydrate	881 μg/mg	Ampicillin, ampicillin sodium, ampicillin trihydrate	
Bacitracin zinc	74 IU/mg	Bacitracin zinc, bacitracin	
Benzylpenicillin potassium	1595 IU/mg	Benzylpenicillin in the form of its potassium, sodium, procaine, benzathine salts, etc.	Penicillins other than benzylpenicillin, esters of benzylpenicillin
Benzylpenicillin sodium	1670 IU/mg		
Candicidin	2098 IU/mg	Candicidin	
Capreomycin sulfate	920 IU/mg	Capreomycin	
Carbenicillin (disodium)	950 μg/mg[b]	Carbenicillin monosodium and disodium salts	
Cephalexin	1000 μg/mg	Cephalexin	Other cephalosporins
Cephaloridine	1000 μg/mg	Cephaloridine	Other cephalosporins
Cephalothin sodium	1000 μg/ml	Cephalothin sodium	Other cephalosporins
Cephradine	1000 μg/mg	Cephradine	Other cephalosporins
Chloramphenicol	1000 μg/mg	Chloramphenicol	Esters of Chloramphenicol
Chlortetracycline hydrochloride	1000 IU/mg	Chlortetracycline	Other tetracyclines
Clindamycin hydrochloride	962 μg/mg	Clindamycin hydrochloride	
Cloxacillin sodium	962 μg/mg	Cloxacillin sodium	
Colistin sulfate	20,500 IU/mg	Colistin sulfate	Colistin sulfomethate
Colistin sulfomethate	12,700 IU/mg	Colistin sulfomethate	Colistin sulfate
Cycloserine	1000 μg/mg	Cycloserine	
Demeclocycline hydrochloride	1000 IU/mg	Demeclocycline hydrochloride	
Dicloxacillin sodium	922 μg/mg[c]	Dicloxacillin	
Dihydrostreptomycin sulfate	820 IU/mg	Dihydrostreptomycin	

Doxycycline hydrochloride hemiethanolate hemihydrate	870 IU/mg	Doxycycline	Other tetracyclines
Erythromycin dihydrate	950 IU/mg	Erythromycin, its ethyl succinate and stearate; also estolate after hydrolysis	
Gentamicin sulfate	641 IU/mg	Gentamicin	
Gramicidin	1000 IU/mg	Gramicidin	
Kanamycin sulfate	812 IU/mg	Kanamycin	
Lincomycin hydrochloride	922 µg/mg	Lincomycin	
Lymecycline	948 IU/mg	Lymecycline	Other tetracyclines
Neomycin sulfate	775 IU/mg	Neomycin	
Nystatin	4855 IU/mg[d]	Nystatin	
Oxytetracycline dihydrate	880 IU/mg	Oxytetracycline, its dihydrate and hydrochloride	Other tetracyclines
Phenethicillin potassium	1000 µg/mg	Phenethicillin	Other penicillins
Phenoxymethylpenicillin potassium	900 µg/mg[e]	Phenoxymethylpenicillin	Other penicillins
Polymyxin B sulfate	8403 IU/mg	Polymyxin B	
Rifampicin	1000 µg/mg	Rifampicin	Rifamycin
Rifamycin sodium	887 IU/mg	Rifamycin	Rifampicin
Streptomycin sulfate	785 IU/mg	Streptomycin	
Tetracycline hydrochloride	982 IU/mg	Tetracycline	Other tetracyclines
Tobramycin	962 U/mg[f]	Tobramycin	
Vancomycin sulfate	1007 IU/mg	Vancomycin, its sulfate and hydrochloride	

[a] Typical potencies are presented as a guide only. In those cases where there is an international reference preparation, its potency is quoted. In other cases a theoretical potency may be quoted by making allowance for hydrates, etc.

[b] Calculated as the disodium salt and assuming that the material contains 5% of sodium benzylpenicillin as an impurity.

[c] Calculated as dicloxacillin acid.

[d] This is the potency of the second international reference standard.

[e] Calculated as phenoxymethylpenicillin acid.

[f] A British reference standard.

Benzylpenicillin in the *Sarcina lutea* agar diffusion assay, 0.05 IU/ml (~0.03 μg/ml).

Doxycycline in the *Bacillus cereus* agar diffusion assay, 0.20 IU/ml (~0.25 μg/ml).

Folic acid (pteroylglutamic acid) in the *Streptococcus faecalis* turbidimetric assay (0.3 ng/ml).

The nature of the problems and means of countering them are discussed in Sections 3.4.1–3.4.4.

3.4.1 Volumetric Glassware

In truly quantitative assays dilutions must always be prepared using volumetric glassware, i.e., bulb pipettes (which would normally be calibrated "to deliver") and volumetric flasks, and perhaps burettes for the preparation of some narrow-range standard curves such as are used in antibiotic turbidimetric assays. However, for liquids other than dilute aqueous solutions a "to-deliver" pipette would not deliver the nominal volume. Thus, for viscous liquids or suspensions as well as for liquids less viscous than water, such as alcoholic solutions, a pipette calibrated "to contain" should be used and rinsed out to remove all the liquid that has been measured. (These pipettes are usually distinguished by the marks D20C, meaning "to deliver at 20°C," and C20C, meaning "to contain at 20°C").

As an alternative to the "to-contain" pipettes, samples may be weighed and the potency calculated first on a weight basis; then, if required, the potency can be converted to a volume basis through determination of weight per milliliter.

The precision of a typical single routine assay is such that confidence limits of ±5% ($p = 0.95$) might be expected.

Table 3.2 shows the respective tolerances permitted by British Standard (BS) Specification numbers 1792 and 1583 for certain sizes of class A and class B volumetric flasks and pipettes. It is clear that, as the tolerances even for class B volumetric glassware are very narrow when compared with the expected assay confidence limits, the use of the more expensive class A glassware cannot be justified in routine work. In contrast, in very precise work such as the 12 × 12 Latin square assays for which confidence limits of ±1% may be achieved, then class A volumetric glassware should be used. However, makers' claims for the quality of glassware cannot be taken for granted. One of the requirements for accreditation by the (British) National Testing Laboratory Accreditation Scheme (NATLAS) is that a laboratory must calibrate all volumetric flasks, burettes, and pipettes. Edmond (1983) reported that of some 1250 volumetric flasks checked for this purpose, 60 (approximately equal numbers of class A and class B) failed to meet the claimed standard.

Referring again to Table 3.2, it will be seen that the smaller the pipette or flask, the wider are the tolerances on a percentage basis. It is recommended, therefore, that for the preparation of test solutions, bulb pipettes smaller than 5 ml and volumetric flasks smaller than 50 ml should be avoided whenever possible.

Table 3.2

Permitted Variation in Capacity of
Volumetric Glassware according to
British Standard Specifications[a]

Nominal capacity	Tolerances			
	Class A		Class B	
(ml)	± ml	± %	± ml	± %
Bulb pipettes, BS 1583				
1	0.007	0.70	0.15	1.50
2	0.01	0.50	0.02	1.00
3	0.015	0.50	0.03	1.00
4	0.015	0.38	0.03	0.75
5	0.015	0.30	0.03	0.60
10	0.02	0.20	0.04	0.40
15	0.025	0.17	0.05	0.33
20	0.03	0.15	0.06	0.30
25	0.03	0.12	0.06	0.24
50	0.04	0.08	0.08	0.16
100	0.06	0.06	0.12	0.12
Flasks, BS 1792				
5	0.02	0.40	0.04	0.80
10	0.02	0.20	0.04	0.40
25	0.03	0.12	0.06	0.24
50	0.05	0.10	0.10	0.20
100	0.08	0.08	0.15	0.15
200	0.15	0.08	0.30	0.15
250	0.15	0.06	0.30	0.12
500	0.25	0.05	0.50	0.10
1000	0.40	0.04	0.80	0.08
2000	0.60	0.03	1.20	0.06

[a] Data reproduced by kind permission
of the British Standards Institution.

3.4.2 Adsorption onto Glass Surfaces

The adsorption of substances onto volumetric glassware is undesirable for two reasons: the accuracy of the volumetric measurement is diminished and active substance is depleted from dilute solutions.

The accuracy that can be attained when using volumetric glassware is dependent on surfaces being wettable. If the inside of the neck of a volumetric flask is greasy, a regular meniscus will not form and so the volume cannot be adjusted properly. In the case of burettes and pipettes calibrated "to deliver," a greasy surface prevents proper drainage and delivery of the correct volume. These effects are proportionately greater when small-volume flasks, pipettes, and burettes are used.

Kavanagh (1982) drew attention to the problem of depletion of active substance through adsorption on the surfaces of containers. This depends on the nature of the surface, its area, the nature of the substance and its concentration, pH, other substances present, and time. The proportion of substance that will be adsorbed from a very dilute solution will be greater than from a stronger solution even though the absolute amount is less. The relatively large surface area: volume ratio in smaller-volume flasks and pipettes is a further reason why these should be avoided, particularly for the lower concentrations in a series of dilutions. Kavanagh (1963) quotes an example of a 2-μg/ml solution of tylosin being depleted to the extent of 23% through adsorption onto the surface of a soft-glass bottle. This is attributed to the cation-exchange properties of the soft glass and the weakly basic nature of tylosin. It may be presumed that the same problem would apply to other basic antibiotics and could be minimized by the use of hard glass and, so far as possible, avoidance of very dilute solutions.

3.4.3 Cleaning of Glassware

To avoid the errors in measurement described in Section 3.4.2, glassware must be kept scrupulously clean and free from grease. The traditional way of cleaning glassware in the chemical laboratory was by soaking a few hours in cleaning acid, a solution of potassium dichromate in strong sulfuric acid, followed by rinsing in tap water and finally distilled water. Although this method is now being superseded (to a large extent on safety grounds) by specially formulated laboratory detergents, it is mentioned here to draw attention to its disadvantages. Chromate ions are adsorbed onto the glass. They are very difficult to remove and they interfere with some microbiological assays. To avoid the problem of chromate ions, Kavanagh (1963) suggests the use of a mixture of 95% concentrated sulfuric acid and 5% concentrated nitric acid. However, the use of either of these highly corrosive mixtures in rather large quantities is a hazard best avoided.

If they must be used, then it is most important to protect the eyes from splashing. Safety goggles or a complete face mask should always be worn. Vincent (1982) reports that for general washing of laboratory glassware, Pyroneg (see Appendix 1) at a concentration of 0.3% is very satisfactory, but for more demanding cleaning such as the removal of traces of vitamins and antibiotics, Decon 90 concentrate (see Appendix 1) as a 0.2–2.0% solution, according to the degree of soiling, is preferred. Vincent found that chromic acid was no better than tap water in removing traces of nicotinic acid adsorbed onto glass. Pyroneg and Decon 90 concentrate were both superior to chromic acid. It is, of course, necessary to rinse very thoroughly after using detergents, as they too may interfere with assays. (See also Chapter 7.)

3.4.4 Sample Preparation and Weighing

To prepare a test solution from a homogeneous solid such as a reference standard or a noncompounded substance, the number of dilution steps, and therefore the errors of dilution, may be minimized by weighing a small amount. However, the quantity weighed should not be so small as to introduce significant weighing errors. For example, an error of 0.1 mg is quite likely when using a four-place balance. Such an error in weighing 100 mg of substance is only 0.1% and is quite acceptable in routine micro-biological assays. An error of 0.1 mg in 10 mg is 1% and, though still small in comparison with the inherent random error of the assay, it is best avoided. It is good policy to weigh at least 50 mg but preferably around 100 mg. When a sample is a compounded solid such as tablets or granules, or is a semisolid such as an ointment or cream, then it is necessary to minimize sampling errors that would arise from heterogeneity of the material. Tablets and granules should be ground to a fine powder with a pestle and mortar and thoroughly mixed before a portion is weighed. The general requirement of the BP 1980 is that 20 tablets should first be weighed, then ground to a powder, then an appropriate quantity of the powder be weighed and used for the assay. Creams and ointments must be thoroughly stirred or mixed before a portion is weighed. Thus, if creams or ointments are packed in collapsible tubes, the entire content of the tube must be emptied and mixed in a suitable container. The mixed material may then be stored in a closed container such as a 60-ml ointment pot for repeat assays if needed.

The weight of a homogenized compounded sample to be taken for assay should depend on the nature of the sample. Thus, for uncoated or film-coated tablets of antibiotics in which the mix might be about 80% of active ingredient and 20% excipients, a weight of 100–200 mg would be suitable. For sugar-coated tablets in which the thickness of coating might vary appreciably, then a higher weight (perhaps 1 g) would be appropriate. In an ointment or cream containing, for example, only 1% of active ingredient (even though it has been homogenized), the risk of sampling errors is greater than in the example of the tablets, and so a weighing of, say, 5 g would seem reasonable. The desirability of increasing sample weight to minimize sampling errors must be balanced against the practical disadvantage of an additional dilution stage being needed, with consequent use of more diluent and more glassware leading to an additional dilution error. The required accuracy of the result and the overall work load on the laboratory must be factors influencing the decision.

For all weighings a balance of appropriate accuracy should be used. A balance reading to 1 mg will suffice for weighings of 1 g or more, whereas a four-place balance (weighing to 0.1 mg) would be suitable for weighings from about 30 mg to 1 g. If circumstances necessitate the weighing of smaller amounts such as 10 mg, then a balance reading to 0.01 mg might be desirable.

Whatever grade of balance is used, it should be checked regularly so as to provide assurance that no significant errors arise in the most fundamental measurement of analytical chemistry — weighing. It is not within the intended scope of this book to give detailed guidance on how to check a balance. It will suffice to say that it is particularly important to ensure that the sensitivity of the balance is correctly set so that, for example, a graticule having a range of 10 mg does actually show between 9.8 and 10.2 mg when a 10-mg weight is placed on the scale pan. It follows that it is necessary to have an authenticated reference set of weights which would be used only for checking balances.

Certain reference standards have to be dried under specified conditions before weighing. Similarly, some samples must be assayed on the "dry basis." Drying requirements (time, temperature, and atmospheric pressure) for a range of antibiotic reference standards are given in the various official compendiums. When drying is necessary a small (~4 cm diameter) squat-form weighing bottle with ground-glass stopper is convenient. First dry the empty weighing bottle in an oven with its stopper tilted to permit escape of adsorbed moisture from the inner surface, then allow to cool in a desiccator containing self-indicating silica gel. Weigh into the bottle just a little more than the required quantity of standard and dry under the prescribed conditions with the lid tilted to permit the escape of moisture. On completion of the drying cycle, allow to cool in a desiccator containing self-indicating silica gel, with the stopper still tilted so that a partial vacuum is not formed on cooling. When cool, remove from the desiccator, close the stopper, weigh the whole, transfer the content via a funnel to a volumetric flask, and reweigh the emptied weighing bottle. When very small quantities are to be weighed, a weighing "pig" is convenient. Small weighing bottles or "pigs" are to be preferred as they present less surface for the adsorption of moisture and so minimize a potential source of error. Moisture adsorption may be still further minimized by not handling the vessels with the fingers directly. Chamois leather gloves may be worn or suitable forceps used to pick up the vessels. An alternative to the use of glass receptacles is to use a "packet" made from heavy-gauge aluminum "kitchen" foil. Such a packet offers several advantages: It is very inexpensive. It is quick and easy to make. Only about 0.1 mg of moisture is adsorbed onto the surface; predrying removes this potential source of error, rendering it negligible. Cooling is so rapid that the packet can be taken straight from the oven to the balance. When drying under reduced pressure is prescribed, it is necessary to ensure that the required vacuum is actually achieved. The gasket on the oven door must be in good condition, and the vacuum pump must be properly maintained. At the end of the drying period, introduce dry air into the oven by allowing it to pass slowly through a tube containing self-indicating silica gel. When drying of the solid material is not required, a weighing funnel is very convenient, because the solid can be washed directly from this receptacle into the volumetric flask.

3.4.5   Preparation of Test Solutions from Samples

The general principles of initial treatment of preparations are the same whether the assay is to be chemical, physical, or microbiological. The methods are designed to obtain a small quantity of homogeneous material that is representative of the whole. For pharmacopoeial products, guidance is given in national pharmacopoeias. The general principles that are expounded in the pharmacopoeias for the treatment of pharmaceutical dosage forms such as tablets, capsules, ampuls, and vials should also be applied to the same dosage forms of products that are not the subject of a pharmacopoeial monograph.

The procedures for certain pharmaceutical forms are illustrated by excerpts from monographs of the BP 1980 that are reproduced here by kind permission of the controller of Her Majesty's Stationery Office.

Capsules

Tetracycline capsules:

To a quantity of the mixed contents of 20 capsules equivalent to 0.25 g of tetracycline hydrochloride, add 500 ml of water, mix, and carry out the biological assay of antibiotics. . . . Calculate the content of tetracycline hydrochloride in the average weight of contents of the capsules, taking each 1000 Units found to be equivalent to 1 mg of tetracycline hydrochloride.

The general statement of the pharmacopoeia on capsules requires that the individual weights of the contents of a number of capsules be determined. This number is normally 20 capsules. The average of these individual weights is the "average weight" referred to in the above excerpt from the British Pharmacopoeia.

Powders for injection

Oxytetracycline injection:

> Determine the weight of content of each of ten containers. . . . Mix the contents of the ten containers and carry out the biological assay of antibiotics, . . . Calculate the content of oxytetracycline in each container, taking each 910 Units found to be equivalent to 1 mg of oxytetracycline hydrochloride.

Tablets

Erythromycin tablets: These are enteric coated (either film or sugar). As the weight of the coating may vary substantially, the individual weights of the coated tablets are not very meaningful and so the pharmacopoeia does not direct that individual weights be determined. (In the pharmaceutical industry the weight variation of the cores before coating may be determined.)

> Weigh and powder 20 tablets. Triturate a quantity of the powder equivalent to 0.4 g of erythromycin with a few ml of sterile phosphate buffer pH 8.0, add sufficient sterile phosphate buffer pH 8.0 to produce 1000 ml and carry out the biological assay for antibiotics, . . . Calculate the content of erythromycin in the tablets taking each 1000 Units found to be equivalent to 1 mg of erythromycin.

It is necessary to mention the "we test what the patient actually receives" school of thought. Following this principle in, for example, the assay of a vial of streptomycin sulfate, the analyst would use a syringe to inject a small volume of water (as described on the label), shake to dissolve the streptomycin sulfate, and then withdraw into the syringe as much as practicable of the resulting solution. The entire content of the syringe would then be diluted to produce the test solutions. The assay would be reported in terms of the number of units of streptomycin activity that could be withdrawn from the vial. This procedure has some serious drawbacks:

(1) Because the weight of the solid content of the vial has not been determined, it is not possible to assign a potency to that solid. There is therefore no way of knowing whether it conformed to pharmacopoeial requirements. A vial that contained the right number of units of activity may have contained an overfill of substandard antibiotic. This is unacceptable.

(2) A repeat assay on a freshly prepared solution from the same vial (i.e., the same weight of material) is not possible, and so there is no basis for assessing the repeatability of assays.

(3) In a repeat assay on the content of another vial from the same batch, assay variation cannot be distinguished from variation in weight of contents.

References

British Pharmacopoeia (1980). British Pharmacopoeia Commission, H.M. Stationery Office, London.

Edmond, J. D. (1983). *Pharm. J.* **230** (6214), 179.

Hewitt, W. (1977). "Microbiological Assay," Chap. 1. Academic Press, New York and London.

Kavanagh, F. W. (1963). "Analytical Microbiology," Vol. 1. Academic Press, New York and London.

Kavanagh, F. W. (1982). Personal communication.

Lightbown, J. W. (1961). *Analyst* **86**, 216.

Miles, A. A. (1952). *In* "Microbial Growth and its Inhibition." World Health Organization Monogr. Ser. No. 10, pp. 131–147.

Vincent, S. (1982). Unpublished data.

Wright, W. W. (1971). *In* "Colloquium of the International Pharm. Res. St., Congr. Pharm. Sci., 31st, Washington, D.C.

CHAPTER 4

THE AGAR DIFFUSION ASSAY

4.1 Essential Features of the Assay

If a dilute aqueous solution is placed in contact with a solid agar gel, solute will diffuse from the solution through the interface into the gel. Ultimately equilibrium will be attained and the solute concentration will be uniform throughout the whole system. During the period of diffusion before equilibrium is attained, concentration of solute in the agar gel decreases with increasing distance from the gel–solution interface. Such a system is used in the microbiological assay of antibiotics by the agar diffusion method, the solute being the antibiotic. The agar gel, which contains nutrients to support the growth and multiplication of microorganisms, is inoculated uniformly with a suspension of a test organism that is sensitive to the antibiotic. Upon incubation, multiplication of the organism commences and continues until it is inhibited by a combination of critical factors. One of these critical factors is the concentration of antibiotic; at all concentrations higher than this "critical concentration," multiplication of the organism ceases. Multiplication continues at the lower concentrations of antibiotic more distant from the reservoir and is eventually revealed by the resultant turbidity of the medium. A clear zone of inhibition contrasts with this turbidity. The position of the zone boundary is dependent *inter alia* on the concentration gradient of the diffusing antibiotic. Zone formation is described in more detail by Hewitt (1977), but for a fuller discussion the reader is referred to the original work reported by Cooper (1963 and 1972).

4.2 Practical Methods of Zone Production

Inhibition zones may be produced in the following ways:

(1) Outward diffusion of the antibiotic from a reservoir into a layer of inoculated agar medium in a dish to yield a circular zone of inhibition.

(2) Downward diffusion of the antibiotic from a solution placed on top of the inoculated agar medium in a vertical narrow tube to yield a cylindrical zone of inhibition.

These two ways are illustrated in Fig. 4.1. The former, commonly referred to as the agar plate method, is by far the most widely used. Some practical techniques are described here.

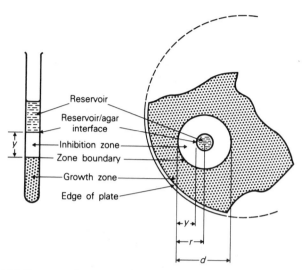

Fig. 4.1.   Inhibition zones in tubes (left) of agar medium as used by Mitchison and Spicer (1949), and plates (right) of agar medium as used in most routine assay procedures. *y*, Distance from reservoir–agar interface to zone boundary (tubes and plates); *r*, zone radius (plates); *d*, zone diameter (plates).

Nutrient agar is melted, then cooled to 50° – 55°C, and inoculated uniformly with a standardized suspension of the test organism. It is then poured into sterilized dishes to form a layer of about 3 mm thickness, and allowed to cool and set. A temperature higher than 55°C might kill or seriously reduce the viable count of a heat-sensitive organism, whereas if the temperature is lower than 50°C, the medium is likely to solidify before it has managed to form a layer of uniform thickness.

The dishes, or plates, as they are commonly termed, may be either petri dishes or square plates (30 cm side), which are usually known as "large plates." Reservoirs may be formed by cutting the agar with a sharp tool resembling a cork borer, and then removing the cylindrical plugs thus formed so as to leave a well in the agar layer. The cylindrical plugs may be removed with the aid of a microspatula or similar tool. As an alternative to wells in the agar, reservoirs may be placed on the surface of the agar. Specially made stainless-steel cylinders are in common use. Unglazed porcelain "fish-spine beads" (normally used as high-temperature insulators for electrical wiring) may also be used. These must be of uniform size, clean, and free from grease. As another alternative, small circular paper disks may be used.

Dilute solutions of the antibiotic, the test solutions for standard and sample described in Chapter 3, are applied to the plate in a manner dependent on the type of reservoir used, thus:

(1)   The solution is added to the wells or cylinders by pipette in accordance with a predetermined pattern and sequence such as described in Section 4.6.3.

(2)   The solution is allowed to soak into fish-spine beads or paper disks, which are then applied to the plate in accordance with the patterns and sequence described in Section 4.6.3.

After a period of diffusion of perhaps 1–2 hours at room temperature or, preferably, in a refrigerator, the plates are incubated overnight at a temperature defined for the particular test organism.

After incubation, clear circular zones of inhibition contrast with the opaque background that is formed through proliferation of the test organism.

In the tube variation of the agar diffusion assay, the inoculated molten medium is dispensed into uniform narrow tubes and allowed to set in the vertical position. Test solutions are added to the tubes, and, after a period of diffusion at room temperature or below, the tubes are incubated overnight. Incubation in a well-stirred water bath ensures uniformity of heating rate and constancy of temperature. These are important factors contributing to the fixation of the position of the zone boundary as explained in Section 4.3. This technique was used by Mitchison and Spicer (1949) for the routine assay of streptomycin. It was chosen by Cooper and Gillespie (1952) for studies on the influence of temperature on zone formation because of the greater control of incubation temperature that could be attained by placing the racks of tubes in a water bath.

4.3   Factors Influencing Zone Size

The factors influencing zone size were studied by several groups of workers mainly in the late 1940s and early 1950s. The studies include those by Cooper and Gillespie mentioned above, Cooper and Woodman (1946), Mitchison and Spicer (1949), Cooper and Linton (1952), Humphrey and Lightbown (1952), and Lees and Tootill (1955a). Their findings were reviewed by Cooper (1963 and 1972).

Briefly, it is propounded that the position of the zone boundary is fixed through the interaction of the following factors:

Critical concentration m': the concentration of antibiotic arriving at the position of the future zone boundary at time t_0.

Critical time t_0: the period of growth of the organism after which it reaches the critical population N'.

Critical population N': the population at time t_0. At higher populations than this the excess of organisms is capable of completely absorbing the antibiotic, thus preventing its further outward diffusion.

These factors are discussed in somewhat more detail by Hewitt (1977), but for a fuller account the reader is directed to the reviews by Cooper (1963 and 1972). Features that are relevant to practical operations in the plate assays are as follows:

(1) *Concentration of the antibiotic in the test solution that is placed in the reservoir:* This is the essential basis of the assay.

(2) *Volume of the test solution in the reservoir:* Provided that the volume is sufficiently great so that the concentration of the antibiotic remains virtually unchanged despite some loss due to diffusion into the agar, then small differences in volume will have little effect. When the reservoir consists of a well cut into the agar layer, then a diameter of about 8 mm provides adequate capacity. It is nevertheless good practice to aim for a constant volume of test solution in each reservoir to minimize any minor variations. When fish-spine beads are used the volume of liquid is small, and so it is essential that the beads be of constant size.

(3) *Density of the inoculum:* A heavier initial inoculum leads to sharper definition of zone boundary. However, it also results in smaller zones and a less steep log dose–response line. It may be noted that larger zones in themselves do not lead to improved assay precision. However, a larger *difference* in zone size over a constant dose ratio (e.g., 2 : 1)—i.e., a steeper slope of the log dose–response line—*does* lead to improved precision.

(4) *Duration and temperature of diffusion phase before incubation:* A period of an hour or more between application of the test solution to the plates and commencement of incubation permits diffusion of the antibiotic before multiplication of the test organism becomes appreciable. A longer period of "prediffusion," as it is termed, leads to larger zones and also to a steeper slope of the response line. Prediffusion may be allowed to take place in a refrigerator, thus minimizing growth during this period but also reducing the rate of diffusion. The best conditions must be found by practical trials. In a single assay there may be a difference of 10–20 minutes between filling the first and the last of the reservoirs. The resulting difference in prediffusion time is a significant factor which must be countered by techniques that will be described later.

(5) *Thickness of the agar medium:* Humphrey and Lightbown (1952) showed that when the reservoirs consisted of small, unglazed ceramic beads

placed on the surface of the agar, a thicker agar layer lead to smaller inhibition zones. Lees and Tootill (1955a) also referred to this same effect.

(6) *Composition of the medium:* A richer medium results in more rapid growth of the organism, in smaller zones, and in a less steep slope of the response line.

(7) *Incubation temperature:* Incubation at any temperature other than the optimum will lead to slower growth of the organism with consequent larger zones and steeper slope of the response line. It is clear that whatever temperature may be chosen for incubation, its uniformity and uniformity of heating rate of the plates is more important than the absolute value of the temperature.

4.4 Dose–Response Relationship

In its early days the plate method was the entirely empirical comparison of responses to an "unknown" with those to a standard reference substance. It was recognized that there were differing responses to the same test solution when the assay was carried out on different occasions. The routine was, therefore, to prepare test solutions at a series of concentrations from the reference standard on each occasion the assay was carried out and to plot the mean response (zone diameter) against dose to produce a standard daily curve. The potency of test solutions prepared from "unknowns" could be estimated from the mean responses they caused, by reading from the standard curve. It was found more convenient, however, to plot mean zone diameters against logarithm of concentration or "dose," as this resulted in a straight line, or at least a line showing only slight curvature. The use of a series of doses forming a geometric progression led to equally spaced doses on a logarithmic scale. This spacing of doses, together with the approximately straight log dose–response relationship made it possible to replace graphic methods of processing observations by purely mathematical methods. Thus, errors of graphic interpolation could be avoided.

The studies on zone formation that were mentioned in Section 4.3 showed on theoretical grounds and confirmed in practice that over a wide range of doses, the square of zone width is directly proportional to logarithm of dose. It should be noted that in both the plate and tube versions of the agar diffusion assay, zone *width* is defined as the distance between the agar gel–solution interface and the zone edge, and must not be confused with the zone diameter, which is normally the observed response in the plate version of the assay. These parameters are illustrated in Fig. 4.1. Clearly, the zone diameter (unsquared) versus log dose relationship cannot also be a straight line. These relationships are illustrated in Figs. 4.2 and 4.3,

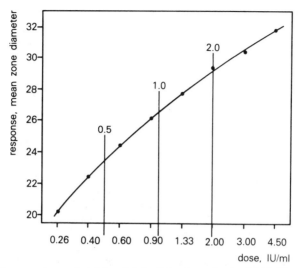

Fig. 4.2. Response to a series of dose levels of streptomycin standard in the agar diffusion assay using *Bacillus subtilis*. The mean of eight zone diameters for each dose level is plotted against the logarithm of the dose. The curvature of the line is evident when viewed over the entire dose range but not so evident over a shorter range such as 4 : 1 or (better still) 2 : 1.

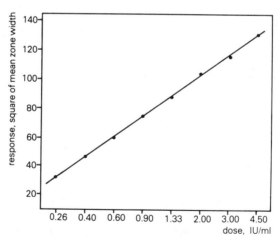

Fig. 4.3. Response to a series of dose levels of streptomycin standard in the agar diffusion assay using *Bacillus subtilis*. The basic data are the same as used in Fig. 4.2, but zone diameters were converted to zone widths by subtracting 9 (the diameter of the reservoir), then dividing by 2. The square of the zone width is plotted against the logarithm of the dose and gives a straight line over the entire range of eight dose levels.

which refer to the same observed data and represent the mean responses on a *Bacillus subtilis* large plate to eight dose levels of streptomycin covering an overall dose range from approximately 0.27 to 4.60 IU/ml. The dose ratio between adjacent dose levels was 3:2 and the overall range was about 17:1, which is much greater than normally used in routine assays. This wide range was used deliberately to demonstrate the curvature of the zone diameter–log dose relationship. Further reference is made to this work in Chapter 8. It will be seen from Fig. 4.2 that if an overall dose range of only 2:1 is used, curvature is unlikely to be apparent. However, if the overall range is increased to 4:1 the possibility of curvature being evident is also increased. In practice, the approximation to a straight-line relationship is quite good enough for most assays; it is convenient and has been found satisfactory over several decades. However, concern as to the validity of assays showing curvature of the response line is indicated by the inclusion of statistical tests in national, regional and the International Pharmacopoeias; these tests show the statistical significance of curvature. This topic is further discussed in Chapter 8.

4.5 Practical Assay Designs

Several alternative shapes and sizes of plates are available for use in the agar diffusion assay, and many different patterns may be used for the arrangement of test solutions made at differing dose levels from the preparations (standard and samples) to be compared. The most commonly used assay plates are probably those listed here.

(1) *Petri dishes of about 9 cm diameter:* These can accommodate 4 or 6 zones having their center points on a circle of radius 2.8 cm. Commonly, each test solution would appear once only on this plate.

(2) *Square plates of side length 30.5 cm:* These can accommodate up to 64 zones arranged in eight rows and eight columns (an 8×8 design). In this case an individual test solution might appear in eight different positions on the same plate.

Other assay plates that have been used include larger petri dishes to accommodate eight or more zones and square plates for 4×4, 6×6, 9×9, and 12×12 designs. Lees and Tootill (1955b) describe the use of rectangular plates for 11×5, 13×4, and 16×6 designs. An example of a 6×6 Latin square design is shown in Fig. 4.4.

The number of dose levels employed is most commonly two or three. One assay may consist of a set of small plates, e.g., six petri dishes giving a replication of six for each test solution; alternatively, it may consist of a

E	D	F	C	A	B
B	F	E	A	C	D
D	B	A	F	E	C
C	A	B	E	D	F
A	C	D	B	F	E
F	E	C	D	B	A

Fig. 4.4. Example of a 6 × 6 Latin square design.

single large plate such as an 8 × 8, in which a replication of 8 is attained within the single plate. However, there are other possible arrangements leading to different replications. A selection of suitable designs can be found in Appendix 8.

A single assay may compare either one sample or a number of samples with one reference standard. A low replication results in a low precision of potency estimate, whereas an assay with a sufficiently high degree of replication might yield a high-precision potency estimate. The underlying logic and relative merits of these various experimental designs is discussed in more detail in Chapter 8. However, for a fuller account the reader is referred to the book by Hewitt (1977).

4.6 The Plate Method in Practice

Irrespective of the substance to be assayed, some general principles and basic techniques are universally applicable. In this section general guidelines are given that will serve as the essential basis enabling the reader to carry out any microbiological assay. By substituting some of the specific requirements such as assay organism and standard reference material, as listed in the individual assay sections, it should be possible to achieve satisfactory results with any of the assays described. It is indeed a good plan to start by experimenting with only one of the assays until sufficient experience has been gained. Then, once one assay is well established and is giving good results, it will be relatively easy to introduce other assays.

4.6.1 Work Planning

The 12 essential steps of the agar diffusion plate assay and their interrelationships are shown in the flowchart of unit operations in Fig. 4.5. Steps 1 and 2, the preparation of nutrient media and buffer solutions, are normally done before the day of the assay, whereas step 3, maintenance of the culture, is a continuous activity. Step 4, preparation of the inoculum, may be done as an overnight culture or up to a few weeks in advance (vegetative forms), or several months in advance of the assay (spore suspensions); step 5, preparation of test solutions, is normally done on the day of the assay, although some relatively stable reference standards may be prepared as stock solutions which may be kept in the refrigerator for a few days or a week, then diluted to dose-level concentrations on each day of the assay. Steps 6–9 (i.e., from preparation of the assay plates to their placement in the incubator) are activities of the first day of the assay; steps 10–12 (measurement of zone diameters, calculation of potency estimate, and disposal of used media, etc.) follow naturally on the second day of the assay. In a busy laboratory, new assays (steps 5–9) are being started on the same day that steps 10–12 of the previous day's assays are being completed. The need for good work planning is clearly evident.

Not a great deal has been written on this topic, because so much depends on personal ability and experience, the scale of operations, and the number of people on the assay team. There is no doubt that much more work can be done with proper planning. The golden rule is to avoid taking on more work than can be completed comfortably within the working day, because too much rushing inevitably will result in work of poor quality. Allow adequate time for the job, especially when first learning the procedures. As more experience is gained, the throughput of samples can be gradually increased.

It will be apparent from the flowchart (Fig. 4.5) that a substantial amount of preparatory work must be completed before an actual assay can be started. The following checklist should serve as a useful reminder:

(a) Unless a previously calibrated suspension is available, subculture the assay organism a day or two before the assay is due, so that a fresh suspension could be prepared in time. Naturally, spore suspensions and cultures for vitamin assays will require a longer period of preparation.

(b) Check that there is sufficient sterile assay medium in stock.

(c) Check that there is a sufficient supply of appropriate buffer solution or other diluent needed.

(d)   Check that there is an up-to-date reference standard.

(e)   Ensure that there is good supply of clean glassware, such as plates, test tubes, sterile cylinders, volumetric flasks and pipettes, sterile graduated pipettes, and disposable polypropylene tips.

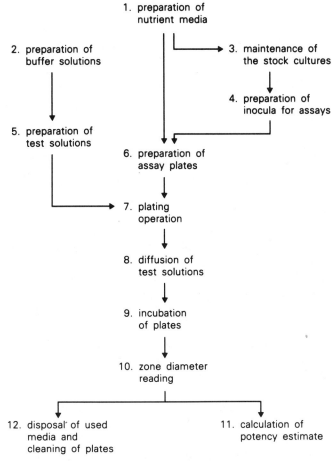

Fig. 4.5. Flowchart showing unit operations in the microbiological diffusion assay.

The simplest situation is when only one type of antibiotic or vitamin needs to be assayed on a regular basis. If more than one type of assay is required, it is better to organize it in such a way that the different assays are done on different days, at least in the beginning. This plan reduces the possibility of committing common mistakes like mixing up the two standard solutions, resulting in a total loss of a day's work. There are, of course, situations in which assaying more than one component in the same sample may be necessary, for example, when more than one antibiotic is present in the same sample, or when dealing with a sample containing a mixture of vitamins. In these cases it is often more sensible to complete the assays on the same extract.

4.6.2   Preparation of Assay Plates

Assuming that the assay medium had been prepared some days or weeks earlier, it will need to be steamed to liquefy the agar. Therefore, the first task of the day is to put the required number of bottles of assay agar into the steamer or autoclave which has a free steaming (melting) cycle. It is advisable to include one or two extra bottles in case of breakages. If screw-capped bottles are used, loosen the caps by about a quarter of a turn before steaming. Do not pack the bottles tightly into wire baskets; this could increase the risk of breakages. Usually the first person arriving to work in the morning is given the job of starting the steaming cycle. Depending on the volume of medium in each bottle, 50–60 minutes of steaming are sufficient to melt the agar. Alternatively, a microwave oven at a low-energy setting may be used.

It is important to ensure that the agar has completely melted and that there are no undissolved lumps left. The easiest way to check for this is to take a bottle of hot agar, tighten the screw cap, and turn the bottle upside down a couple of times and examine the liquid against the light. (Wear heat-insulated gloves.) If the agar has melted properly, air bubbles will be seen to rise unhindered through the liquid. If the air bubbles do not rise in a straight line but seem to go around some invisible objects, then the agar is lumpy, i.e., it is not fully dissolved and needs more heating. In this case, carefully unscrew the cap just enough to release any pressure that may have built up and replace the bottle in the steamer for an additional 10 minutes or so.

It is reemphasized that overheating the medium can impair its nutritive properties as well as weaken its gelling strength, especially if the medium happens to have a low pH. A useful addition to the equipment list is a simple pocket timer or alarm clock, which can be set to provide an audible warning when the heating cycle should be terminated. The more modern but expensive microprocessor-controlled autoclaves can be programmed to turn off the steam automatically when the heating cycle is completed.

Another important task that needs to be done first thing in the morning is to switch on the water bath so that the water reaches the required temperature by the time the agar medium has melted. The temperature of the water bath is commonly controlled at about 50°C, but sometimes a lower setting is preferred when, for example, a particularly heat-sensitive organism, such as some strains of *Escherichia coli* or the lactobacilli used in the vitamin assays, is used for inoculum. On the other hand, it is sometimes an advantage to hold the agar at a higher temperature ($55° - 60°C$), when using a bacterial spore suspension to inoculate the medium.

It is worth bearing in mind that agars of different origin may have different gelling properties; e.g., some Japanese agars tend to set at higher temperature than New Zealand agar. Refer to Chapter 7 for further details about testing of agar for suitability in diffusion assays.

Do not plunge the hot bottles of agar medium into the water bath as soon as they are taken out of the steamer, unless heat-resistant glass bottles are used, but stand them on a clean towel or several layers of paper tissue for about 15 minutes to cool. If the bottles are placed directly on a cold bench or other cold surface, some of the bottles might crack or the agar might start to set in the bottom layer. This could also happen if the bottles are left out of the water bath for too long and the agar becomes "lumpy." Check the temperature of the medium from time to time using a mercury–glass thermometer. The nominal $0° - 100°C$ range is quite suitable.

It is not necessary to use aseptic technique or to use a sterile thermometer for this purpose. (An explanation will be supplied later in the chapter.) When the temperature of the medium has dropped to about $65°C$, immerse the bottles in the water bath until they are needed. The water level in the bath should cover the top of the agar in the bottles, but make sure that the bottles do not float or topple over, because the water in the bath can be a source of heavy contamination (see also Chapter 7). If conical flasks are used for the medium, it may be necessary to fit ring-shaped lead weights to weigh them down.

While the medium is cooling down some further preparatory activities can take place, such as setting out the required number of plates and getting them leveled. If petri dishes are used, it is desirable to install an accurately leveled plate-pouring benchtop, preferably made of stainless steel or marble or some similar material that is cold to the touch. Lay out the petri dishes in neat rows and columns after marking each one with indelible ink. Mark the back of the plates so that if the lids get knocked off accidentally, the unique identifications are retained. The plates can be simply marked consecutively and a record kept on a worksheet or in a laboratory notebook. Abbreviations can be used (e.g., PEN for penicillin assay, NEO for neomycin assay). If more than one assistant is involved, their initials need to be indicated. Finally, a mark is necessary to indicate the starting position (e.g., the zone at position 12 o'clock) corresponding to the solution which was first dispensed according to a design or template, the use of which is essential to avoid anticipation of zone diameters when "reading," i.e., measuring. The rest of the solutions are plated out in a clockwise direction, each position being marked with the appropriate solution number. Suitable designs are described by Hewitt (1977).

If large plates are used, they are usually leveled individually by means of special leveling stands, one for each plate. These are triangular in shape and are fitted with adjustable legs. The large plate is assembled, if it is the knockdown type which are popular because they are easy to clean, and the glass base is placed on top of the leveling frame. Figure 4.6 shows a typical leveling frame, an assembled large plate without lid, and a simple hand-held agar cutter. To level the plate, take off the lid and place two spirit levels at right angles inside the plate. Adjust the two leveling screws (legs) at the front of the leveling stand until the air bubble in both spirit levels stays in the center. Change the two spirit levels around to check that the plate is truly level. Replace the lid and mark the top of the lid with the abbreviated name of the substance being assayed, the initials of the assistant carrying out the assay, and the plate number of the assistant (e.g., PEN/LMA6 or NEO/YF3).

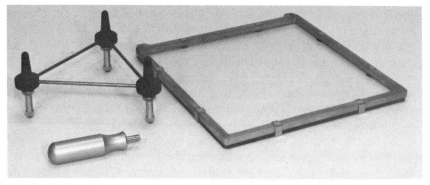

Fig. 4.6. A typical 30 × 30 cm large plate with detachable aluminum frame, which is held onto the sheet of glass by means of eight spring clips. A triangular leveling stand and a hand-held agar-cutting tool are also shown.

To ensure good reproducibility, check the pH of each bottle of medium and adjust it if necessary to the value recommended under the individual assay section (Chapter 9).

Strips of pH paper may be used, provided that the required pH value lies in the middle of the range on the pH strip. However, for accurate measurement a pH meter, which has to be fitted with a special electrode that is suitable for measuring the pH of hot agar solution, is preferred. For further details refer to Section 7.11.4.

Most assay media are formulated to have a predetermined pH value after sterilization, but the pH sometimes changes, especially if new bottles are used for the first time. Indeed, it is often more convenient to prepare some of the media without any pH adjustment because the same composition of medium can be used in different assays by changing the pH. See Appendix 4 for examples.

It is convenient to use 3 *M* hydrochloric acid to lower the pH and 3 *M* sodium hydroxide to raise it. Add the appropriate solution to the medium one drop at a time; tighten the screw cap and mix the contents by gently rolling and tumbling the bottle over three or four times. Do not shake the bottle, as any foam formed will be difficult to disperse.

Carefully unscrew the cap and check the pH again. Add more of the appropriate solution dropwise in this manner until the desired pH value is attained. Keep a record of the amount of adjusting fluid needed to adjust the pH of the first bottle, after which the rest of the bottles can be adjusted rapidly on the same day. Check and adjust the pH of each bottle of medium and keep a record of pH found, pH adjusted to, and the amount of adjusting fluid added on each occasion.

Any other ingredients that may be required can be added at this point, e.g., dextrose solution, which is usually prepared separately from the medium as a 40% (w/v) sterile solution.

If the inoculum to be added is a spore suspension, this should not present any problems. The spore suspension should have been calibrated on a previous occasion as described in Section 2.8, and by now should bear on the label the name and reference number of the organism, the date of calibration, the date of expiry, and the volume to be used per set volume of medium. The assay medium also should be available sterile in convenient quantities and should have been tested soon after it was prepared. When the medium is to be used for petri dish assays a convenient volume would be whatever is easily handled (e.g., 100 or 200 ml).

For large plates (~ 30 cm square) a convenient volume would be that which is just sufficient to fill one plate. In general a volume of about 240 ml should be satisfactory. The volume of inoculum is usually 0.5 – 1.0 ml and does not normally exceed 1% of the volume of the assay medium. Too large a volume of inoculum could dilute the medium too much and could affect the gel strength of the agar. Inocula used in vitamin diffusion assays often greatly exceed the 1% figure quoted above, but in these cases the volume of double-strength medium is suitably adjusted to take this into account.

When the seeding inoculum is in the form of a spore suspension, the temperature of the medium may be allowed to be as high as 60° – 70°C at the time of adding the inoculum, in which case heat-resistant glass plates should be used. In addition to the pasteurization effect described in Chapter

2, the extra heat shock tends to stimulate synchronized germination of the spores, which should result in more uniform background growth and better zone definition.

On the other hand, heat-sensitive vegetative bacterial suspensions must not be exposed to the hot medium any longer than absolutely necessary. For this reason the medium has to be cooled to the lowest possible "safe" temperature, i.e., just before the agar begins to solidify.

If there are a lot of bottles to be cooled, the water bath should be set at the desired temperature. If only a few bottles of media are involved, take the bottles out of the 50°C water bath and place them on several layers of absorbent cloth such as a folded towel or dish towel. Check the medium every 5 minutes or so with a mercury–glass thermometer until the required temperature is reached. Remember that agars of different origin may have different setting temperatures; e.g., Davis New Zealand agar sets at about 47°–48°C.

The question has often been asked whether aseptic techniques were necessary to carry out microbiological assays or whether laminar-flow cabinets were needed. Aseptic techniques are necessary for culture maintenance and for keeping the stock culture and stock spore suspension free from external contamination. Therefore, the emphasis is on protecting the integrity of the culture suspension because it is usually needed for the inoculation of several bottles of media, possibly over several days or in the case of spore suspensions maybe over several months. It follows, therefore, that sterile graduated pipettes must be used to inoculate the assay medium, but apart from sound aseptic techniques laminar-flow protection should not be necessary.

It was mentioned earlier that it is not necessary to sterilize the thermometer when checking the temperature of the medium before inoculation and that it was not necessary to use sterile plates either. Standard and sample solutions are not diluted using aseptic techniques; volumetric glassware is not normally sterile nor are the diluents. Spotting out of test solutions is again not an aseptic operation. It is obvious, therefore, that laminar-flow protection is not necessary.

If it is appreciated that the average number of colony-forming units (CFU) introduced when the seed layer is inoculated is in the order of $10^8 - 10^9$, then it will be easy to see that any external contamination of less than 100 CFU will have little chance of proliferation in competition with the massive number of assay organisms in the short period of overnight incubation.

Even though laminar-flow protection should not be necessary, this does not mean that careless and sloppy techniques could be allowed. On the contrary, all reasonable precautions must be taken to establish and maintain as high standards as possible. Strict attention to small details is necessary at all times.

It is worth mentioning at this stage of the procedures that water baths can be responsible for significant contamination during microbiological assays. Culture media can find their way into the water either from accidental spillage or simply from traces on the outside of the bottles, and thermoduric organisms can proliferate in the water. Since millions of these organisms could be present, the introduction of just a drop of water into the bottle of assay medium can result in a serious contamination.

This problem can be avoided in two possible ways. First, the water bath has to be kept scrupulously clean at all times by scrubbing the bath at the first sign of turbidity or sliminess and by changing the water regularly. Second, it is good practice to wipe each bottle dry with a clean cloth when the bottle is taken out of the bath for pH determination or inoculation. It is especially important to make sure that the bottles are dry when the seeded agar is poured into the assay plates. This is the most likely occasion when a drop of contaminated water from the outside of the bottle could run into the assay dish and could get mixed with the seeded agar.

Add the appropriate inoculum to the molten and cooled medium as indicated in the individual assay section (Chapter 9) using a sterile graduated pipette, then discard the used pipette into an autoclavable pipette jar which contains a suitable disinfectant solution. With the aid of a safety device, withdraw about 1 ml more of the microbial suspension than is needed to inoculate all the bottles of media. The same pipette can be used for all the media requiring the same suspension, but do not use the last 1 ml or so because the tip of the pipette is usually the least accurate. Any suspension left in the pipette after inoculation should be discarded into the disinfectant solution; on no account should it be returned to the stock suspension. If several bottles have to be inoculated at the same time, select a pipette which is large enough for the purpose (e.g., 5 or 10 ml), because it is important to reduce the number of occasions to a minimum when the stock suspension could be contaminated. For example, to inoculate four bottles of assay medium fill a 5-ml sterile graduated pipette to the mark with the microbial suspension; dispense 1 ml into each of the four bottles and discard the pipette, together with the residual 1 ml of suspension, into the pipette jar.

After adding the inoculum, tighten the screw cap and gently mix the inoculum with the medium by rolling and tumbling the inoculated bottle, taking care to avoid the formation of air bubbles as far as possible, and pour the medium immediately into the previously leveled plates. When large plates are used, pour the seeded medium from the bottle into the large plate while moving the bottle in a circle over the plate, to ensure that the whole surface is covered in a uniform layer of about 3 mm thickness before the agar has a chance to solidify. This obviously needs some practice, but the process is reminiscent of making pancakes.

Adding inoculated assay medium to petri dishes may be done by using sterile 12.5-ml or 25-ml rapid-flow pipettes. Old pipettes with damaged tips can also be used for this purpose after carefully cutting off the damaged tip. Exact volumes are not essential so long as the volumes dispensed are the same in each dish. Have the medium available in convenient volumes, such as about 130 ml, which should be sufficient for 10 petri dishes, or about 260 ml, which should be sufficient for 20 plates. Bear in mind that the media should be dispensed into the petri dishes as rapidly as possible, especially if heat-sensitive vegetative organisms are used.

After pouring the inoculated agar into the plates, burst any air bubbles which rise to the surface with the aid of a Bunsen flame by passing the flame momentarily over the air bubbles before the agar has had time to set. Do not overdo the flaming, because this could set up areas of localized dehydration or could even destroy some of the inoculum thus causing the formation of irregular zones. Even more care is needed when plastic petri dishes are used, because excessive flaming could melt the plastic and distort the dishes.

After pouring the inoculated medium into the large plate, flame the underside of the aluminum lid and put it back on the plate in a slightly raised position to allow excess moisture to evaporate. The lid can be supported on three or four rubber bungs placed around the top of the frame with the lid resting on the rubber bungs. If petri dishes are used, pull the lid over the top of the base plate slightly so that a small gap is left for the moisture to escape without exposing more of the agar surface than absolutely necessary. This should prevent the formation of too much condensation on the agar surface.

Leave the plates in this position for not longer than 20 or 30 minutes, then replace the lids and stack the plates upside down in a refrigerator until they are required for "plating out." The purpose of this step is twofold:

(1)   The colder temperature helps to ensure a firmer gel, especially if the agar is going to be cut and the cut-out agar disks have to be removed for the "cup–plate" assay.

(2)   If the agar is cold when the test solutions are applied, the active ingredient will diffuse farther before the assay organism begins to grow, thus producing larger and better definition of zones.

This technique is especially useful when assaying slowly diffusing compounds and often makes prediffusion unnecessary.

4.6.3   Application of Test Solutions to the Assay Plates

It was indicated in Section 4.2 that test solutions can be applied to inoculated assay media in petri dishes or large plates in a variety of ways, such as by "fish-spines," paper disks, cylinders, or wells cut in the agar. These methods will be discussed in more detail in the following paragraphs.

Preparation of standard and sample solutions has been treated fully in Chapter 3. Some of the standards may need to be dried for up to 3 hours; therefore, it follows that weighing of standards and samples must begin fairly early in the morning so as to allow sufficient time for drying, dissolving, and further dilution to the two or three different dose levels. The secret of good organization is to fit the standard and sample preparation into the working day in between all the other preparatory work. Once again, planning and timing the various steps are of paramount importance.

Every flask used in the preparation of test solutions must be clearly marked using either a suitable peelable label or indelible ink directly on the glass surface. Self-explanatory abbreviations should be used (e.g., NEO STD for neomycin standard solution). In addition, indicate the actual or nominal concentration (e.g., 1000 μg/ml or 20 IU/ml).

Naturally, flasks containing sample dilutions must bear the appropriate sample reference number. The weighing and dilution steps must be fully documented to comply with good laboratory practices. Chapter 9, which describes individual assays, contains numerous examples in terms of recommended dilution steps as well as the final dose levels, so as to make this process as clear as possible.

When the final plating-out solutions have been prepared, transfer them from the volumetric flasks to appropriately labeled suitable containers so as to facilitate their application to the assay plates. Depending on the chosen method of application, the standard and test solutions may be poured into 50-ml beakers if paper disks or beads are used, or into test tubes in the case of cup–plate assays. At this stage each solution must be given a serial number starting with the number 1, and these numbers have to be written on the sides of the beakers or on the racks for each position that the tubes occupy. For example, solution numbers 1–6 would be used in a typical petri dish assay, or numbers 1–8 in a typical 8×8 Latin square assay. Before plating out the solutions, a template is placed under the assay plate which contains the inoculated assay medium. The random numbers, which correspond to the test solutions, can be seen clearly through the agar layer. The numbers on the template indicate which solution has to be applied to which position on the assay plate. Mark the back of the petri dish where the first solution is to be placed. The rest of the solutions then follow in a clockwise direction. Full details will be given later. Depending on the method that will be used for applying the solutions to the plates, there are some further preliminary preparations to be completed.

Fish-spine beads.   One of the simplest but currently little-used methods is to employ the so-called fish-spine beads. These are unglazed porcelain insulating beads, normally used for electrical heating elements; they can be obtained in No. 2 or No. 3 sizes, usually by weight, from main electrical suppliers. When the beads are first received, they have to be examined and any chipped or cracked beads discarded; occasional small burrs present on the domed end can be easily removed with the thumbnail. Lees and Tootill (1955a) have fully evaluated the use of these beads and they recommend that the beads should be treated before they are first used and after each assay by boiling in 50% (v/v) hydrochloric acid for 10 minutes, followed by washing with water until the washings are free from acid, and finally by heating in a muffle furnace at $500° - 700°C$ for up to 2 hours. This procedure ensures that the beads are free from grease, which is necessary for a constant take-up of test solutions. Any beads which are not used immediately can be stored if they are protected from dust, but they must be sterilized in a hot-air oven (e.g., 160°C for 2 hours) after counting out the numbers needed for the next assay.

Fill 50-ml beakers, which have been marked with the number of the appropriate test solution, right up to the brim to facilitate filling the beads. Using sterile forceps, pick up one of the beads, holding it with the domed end uppermost, and touch the surface of the first solution to be plated out with the base of the bead. Do not immerse the bead, and try to keep it as vertical as possible. The central hole of the bead will fill automatically with liquid by capillary action. Place the bead just filled — still holding it with the domed end uppermost — onto the surface of the seeded and solidified assay agar over the first position as indicated by the design. Carry on this way with each bead and each solution, following the design until all the places have been occupied. Petri dishes or large plates are equally suitable for this method.

After application of the beads, carefully transfer the plates to the appropriate incubator, having replaced the lids. Reading of the zone diameters will proceed in the morning in the usual manner. Figure 4.7 shows a petri dish with fish-spines after incubation.

Paper disks.   The European Pharmacopoeia and the British Pharmacopoeia (1980) allow the use of paper disks without giving instructions as to their proper use. The size of the zone can be greatly affected by the thickness and composition of the filter paper. For this reason, ordinary filter paper disks cut out with a cork borer are not suitable. Filter paper manufacturers (e.g., Whatman) market assay disks of uniform thickness and diameter

Fig. 4.7.   Clear zones of inhibition around fish-spines in a glass petri dish after incubation.

which are specially made for this purpose. Before they can be used, they have to be placed inside a suitable container such as a glass petri dish and sterilized in an autoclave at 121°C for 15 minutes.

The following procedure is acceptable for routine control assays where a precision of ±5% is sufficient, e.g., for monitoring production batches. Using sterile forceps, pick up one of the sterile paper disks and dip it into the first solution to be plated out, making sure not to touch the liquid with the forceps. Touch the disk on the edge of the beaker to remove excess solution from it, and place it immediately onto the agar surface so as to coincide with the first corresponding position on the random design showing through the medium. Continue in this manner using new paper disks until all the positions have been filled. Replace the lid and incubate the assay plate overnight. Proceed with the measurement of the zone diameters the following morning as usual.

For assays of higher precision, the following procedure may be recommended. Using an appropriate statistical design as a guide, place a sterilized paper disk with sterile forceps on the surface of the seeded assay agar, taking care not to trap any air under the disk. One by one, place the assay disks just above the treatment numbers, which can be seen through the layer of assay medium. The agar surface must be reasonably dry so that the paper does not get saturated with moisture drawn from the medium. Transfer accurate volumes of the test solutions with the aid of a micropipette or microsyringe onto the disks one by one, following an appropriate design. Pipetters with disposable polypropylene tips, such as the ones described in the cup–plate assays later on, can also be utilized. Some workers prefer to put the paper disks on a clear sheet of glass, with the design underneath the glass, and charge the disks with the test solutions. After the paper has dried, the paper disks are transferred to their respective positions on the agar surface.

Note: It must be emphasized that commercially available paper disks which have been impregnated with specified concentrations of antibiotics, the so-called sensitivity disks, are intended for use in clinical microbiological laboratories to test the sensitivity of cultures isolated from patients against the named antibiotics. Such disks must not be used as reference standards as they are quite unsuitable for this purpose.

Cylinders. The cylinder–plate method is widely used in the United States and is described both in the USP and the CFR. The cylinders are made of either steel or porcelain and should have the following dimensions: length 10 mm, inside diameter 6 mm, and outside diameter 8 mm. Each measurement has a tolerance of ±0.1 mm. A satisfactory cleaning regime of the cylinders has been detailed by Hans *et al.* (1963). From time to time the cylinders have to be boiled in alcoholic sodium hydroxide solution, followed by rinsing in tap and distilled water until a neutral reaction of the water is obtained; finally, the cylinders are rinsed with alcohol and sterilized inside a glass petri dish in a hot-air oven.

For the assay the cylinders should be dropped onto the agar surface by means of a special applicator from a height of 12 mm. This ensures that the base of the cylinders sink just below the surface, thus preventing the solution from leaking onto the surface. The cylinders should be placed in a regular pattern and equidistant from one another on a radius of about 2.8 cm from the center of the plate. Six cylinders can be accommodated this way on a standard petri dish.

Wallhäußer (1982) points out that the cylinder technique can be used only with glass petri dishes because of the height of the cylinders. He also describes some specially designed plastic dishes which have the cylinders built into the lids. The cylinders cannot be used on large plates, because with 64 cylinders on a plate they could be easily knocked over during plating out and the lids could not be replaced on the plate. One advantage of cylinders is that the test solutions need not be sterile; therefore, fermentation liquors and tissue extracts can be assayed by this method.

Filling cylinders with test solutions should follow the same statistical designs as in the previous techniques. Filling the cylinders with solutions up to the rim is not a satisfactory method if accurate results are required. It is far better to use a micropipette or some similar device which can deliver an accurate and uniform volume of liquid to each cylinder. Great care is necessary when the petri dishes are transferred to the incubator to make sure that the solutions are not spilled and that the cylinders are not knocked over. After overnight incubation the zone diameters are best measured with an optical system such as the Fischer–Lilly zone reader. The cylinder–plate method has not gained popularity in Britain, mainly because of the wide-spread use of large plates. It was formerly assumed that the cylinder–plate method was more sensitive than other methods because of the relatively large volume of test solution in the cylinders. However, in the authors' experience this does not appear to be the case.

Cup–plate assays.   The most popular method, certainly in Great Britain, is what is generally referred to as the cup–plate method. A tool similar to a cork borer is used to cut out circular cylindrical disks from the seeded agar medium. This agar-cutting tool differs from the cork borer in one important detail; namely, it is sharpened from the inside, whereas the cork borer is sharpened from the outside. The reason for the difference is that if a cork borer were to be used to cut a hole, the agar would be forced away from the edge of the cork borer with the consequent danger of splitting the agar, which would almost certainly result in a misshapen zone. The specially designed agar cutter, on the other hand, being sharpened on the inside edge, upon insertion into the agar forces the agar inward, exerting pressure on the agar plug, which is discarded anyway. A suitable template should be used to ensure that the agar is cut in correct symmetric positions.

Some laboratories use a homemade device which cuts and lifts the agar disk simultaneously, usually into a vacuum chamber. These gadgets tend to work well only in experienced hands, because there is always the danger that the vacuum will lift the agar layer in the plate, allowing test solution to leak under the agar and thereby causing distorted zones.

Before the introduction of large plates some 40 years ago in Great Britain, laboratories used petri dishes with only four cavities cut in the agar. There is still at least one government laboratory using the same system, but the whole procedure has now been fully automated. It is possible to cut up to six holes or cavities in the agar. When using petri dishes it is important to ensure that the dishes have flat bases, so that when the agar is poured into them a layer of uniform thickness forms. Some plastic dishes have a slightly convex base, resulting in a thinner layer of agar in the middle; this, in turn, will result in irregular, noncircular zones. It is essential to dispense the solutions following a suitable statistical design. Hewitt (1977) gives examples of satisfactory designs suitable for petri dishes.

The wells can be cut one by one using a hand-held agar-cutting tool, although it is difficult to ensure that the cutter enters the surface at right angles every time. A hand-held single agar cutter is shown in Fig. 4.6. There are commercially available instruments which will cut all the required positions on the same plate simultaneously, whether it is a petri dish or a large plate. For example, Scientific and Technical Supplies Ltd. (Landwades Business Park, Kennett, Nr. Newmarket, Suffolk, CB8 7PU, England; formerly known as Autodata) markets a 64-place agar cutter suitable for large plates. The instrument is illustrated in Fig. 4.8. The same firm also supplies the large plates with demountable aluminum frames, which are fitted with locating points to ensure accurate alignment on the 64-place agar cutter.

Fig. 4.8. A 64-place (8 × 8 pattern) Token agar plate punch, having just cut through the inoculated agar layer within a 30 × 30 cm large plate.

The single hand-held cutter may be kept conveniently in a small beaker or a suitable screw-capped bottle containing cotton wool at the bottom to protect the cutting edge and filled with methylated spirit (IMS) or 2-propanol (isopropyl alcohol, IPA). A shallow tray which fits under the 64-place cutter can be filled to a depth of about 1 cm with IMS or IPA, so that the spring-loaded cutters can be lowered into it both before and after use. If more than one assay organism is used on the same day, the following procedure is recommended.

Sort the plates which have been inoculated with different organisms into separate groups, and proceed to cut through the agar layer in each plate belonging to the first group. Before starting on the next group of plates which contain a different assay organism, place the shallow dish with the alcohol under the set of cutters and dip the cutters repeatedly into the alcohol. Set the tray with the alcohol to one side, and carry on with the second group of plates. Finally, finish with rinsing the cutters in the alcohol after the last plate.

Remember when handling alcohol to extinguish any naked flames, including pilot lights on Bunsen burners, and pour residual alcohol into a flameproof safety can for safe disposal later on when the day's agar-cutting operation is finished. Refer to Section 7.11.3 for cleaning and maintenance of the cutters.

Having cut through the agar layer, it will now be necessary to remove the cylindrical plugs of agar. The authors have seen many homemade gadgets for this purpose, such as paper clips, hairpins, pen nibs, thin spatulas, and dissecting needles, the tips of which have been bent or hooked to facilitate lifting out the plugs. The removed agar plugs should be dropped into a small dish containing a suitable disinfectant solution, such as 2% (v/v) aqueous Hycolin, and should eventually be autoclaved at 121°C or higher for 30 minutes, when the hot liquid can be safely flushed down the drain and the container can be washed and dried, ready for the next occasion.

The tools favored by the assay staff for "picking," i.e., for lifting out the cut-out agar plugs, should be kept in a small pot or a suitable stand so that the sharp ends are upside down out of harm's way. Flame the tips of these gadgets until they are glowing red before and after use; alternatively, they can be stored in 75% (v/v) aqueous IMS or IPA.

The agar-"picking" tool favored in the author's (Vincent) laboratory has always been a mounted dissecting needle with its tip bent at right angles to a length of about 2 mm. With experience a row of eight plugs can be picked up on the same tool at the same time by means of a quick stabbing and flicking movement of the wrist.

When all the agar plugs have been removed from the plate, replace the lid and transfer the plate upside down into the refrigerator until required for "plating out."

When it is time to dispense the standard and test solutions into the cavities, called "cups," take out one of the plates from the refrigerator and place it carefully on the bench. Turn the plate the right way up, leaving the lid on the bench. Place the plate over one of the statistical designs taken at random, so that the treatment numbers show through the agar layer.

Position the plate so that the numbers are just below the cups. Refer to Appendix 8 for suitable designs. Using clean and dry paper tissue or absorbent cotton wool, wipe off the condensation from the lid (if present) and replace the lid on the plate.

Solutions were formerly dispensed into the cups by means of "dropping pipettes," which were made of glass tubes with a narrow platinum tube sealed in one end and a rubber teat fitted at the other. The first solution indicated by the design was withdrawn into the dropper by means of the rubber teat, and four drops of the solution were counted into the first cup. The residual solution in the dropper was discarded into a waste container and the dropper was rinsed three times with plain buffer solution, each time discarding the rinsings into the waste container. The dropper was rinsed three times with the next solution indicated by the design, and the rinsings were discarded before four drops of this solution were counted into the next cup. This process was repeated until all the solutions were dispensed.

Rinsing the dropper between each treatment was important to avoid carry-over from one solution to the other, especially from a high concentration to a low one. In those days it could take up to 20 minutes to fill all 64 cups on a large plate. This could introduce a significant time factor affecting the development of zone diameters. This method of dispensing was described by Simpson (1963).

A fixed-volume dispenser was introduced by Vincent (1970) to replace the platinum-tipped droppers. A 100-μl "Oxford" dispenser made by Oxford Laboratories (107 North Bayshore Blvd., San Mateo, California 94401) was found to be satisfactory to fill cups which were cut with an 8.5-mm internal diameter cutter and an agar layer of about 3 mm thickness. There are many other makes of dispensers on the market, and guidance on how to test them is given in Section 7.10. The essential feature of these dispensers is that the dispensed liquid does not come in contact with the working parts of the instrument, because nonwettable polypropylene tips are used to pick up the solutions. To use the dispensers properly, follow the manufacturer's instructions.

As a general guideline the following procedure can be recommended:

(a) Fit an appropriate tip to the cone of the dispenser.

(b) Immerse the end of the tip just below the surface of the first solution to be plated out and depress the plunger. For ease of control rest the tip gently against the side of the container.

(c) Release the plunger, drawing liquid into the tip. Withdraw the dispenser, sliding the point of the tip along the inside wall of the container. This action will remove any excess liquid clinging to the outside of the tip.

(d) Dispense the contents of the tip into the first cup by depressing the plunger fully. Take care not to splash the solution outside the cup. Beginners may find it easier to rest the point of the tip gently on the glass surface inside the cup while dispensing the solution into the cup.

(e)   Repeat the above steps for each cup, using the same tip and following the design from left to right. Note that the same tip is used to dispense all the solutions on the same large plate without rinsing between the solutions. Only one tip would be needed to dispense all the solutions into a set of petri dishes also.

(f)   When all the cups have been filled, discard the tip and fit a new tip for the next set of solutions.

Figure 4.9 shows one of these dispensers in use.

Visitors to the author's (Vincent) laboratory were often incredulous that solutions could be dispensed with the same tip without rinsing in between solutions. Manufacturers claim that the tips are nonwettable; therefore, by implication, do not carry over any liquid from one solution to the other. In practice traces of solutions may be carried over, but this is normally well below the threshold of sensitivity. There is one sure way to prove the point beyond all doubt: that is to use the test method described in Section 7.10. There are exceptional circumstances when even the traces carried over by the tips could be significant. For example, separate tips must be used for each solution which contains enzymes such as penicillinase.

4.6.4   Diffusion and Incubation

Having filled all the cups with the appropriate solutions, it is now necessary to transfer the plates to the incubator. Carry the plates carefully, holding them level, and slide them inside the shelf of the incubator without knocking or spilling the solutions. If petri dishes are used, do not stack more than four plates on top of one another. Incubator shelves should be perforated to allow free circulation of air so as to maintain uniform temperature around the plates. If large plates are used, they should be on separate shelves. A space of about 7 cm between shelves is usually adequate. When using large plates of the Autodata design, it was found that the spring clips tended to get caught on the perforated shelves as the plates were being pushed inside the incubator. The problem was overcome by constructing special open shelving made from stout stainless-steel wire in the shape of a square frame which could be fitted onto the shelf supports. Parallel bars running toward the back of the incubator were soldered on top of the frame, so that the plates were lifted just enough to clear the spring clips as the plates were pushed onto the shelves. An even simpler solution is to place two square-shaped bars, made of stainless steel or aluminum, having a cross section of about 1 cm^2, about parallel and both pointing toward the back of the incubator on one of the shelves. Put a large plate on top of these bars,

Fig. 4.9. A typical sampler/pipetter, fitted with a nonwettable polypropylene tip which is used for transferring constant volumes of test solutions into random positions on a large plate in the so-called cup–plate assay.

then two more bars on top of the lid, making sure that the bars are long enough to rest on the front and back of the metal frame and do not press down on the center of the lid. It is now possible to put another large plate on top of these bars, thus leaving a 1-cm gap between the plates. It is not recommended to stack more than three or four plates in this manner.

4.6.5 Measurement of Zone Diameter

After overnight incubation at the recommended temperature, the zones should have developed sufficiently to allow their accurate measurement. This is perhaps the most critical part of the assay procedure because both the potency and the confidence limit calculations depend on these measurements.

The simplest method is to measure the zone diameters with hand-held vernier calipers. (*Note:* not with a plastic ruler!) Pharmacopoeias and the literature state that each zone diameter should be read to the nearest 0.1 mm. After many years of experience it can be stated that it is easier said than done. It should be realized also that some people are more suitable for this task than others. This topic is further discussed in Section 7.13.

One of the least expensive and at the same time one of the easiest zone-measuring instrument is the "clock-dial" caliper, the jaws of which have been ground into needle points. Figure 4.10 shows such an instrument. A common situation is that one of the assistants makes the measurements by lining up the needle points of the caliper with the widest points of each zone and dictates these to a second assistant who writes them down.

Surprisingly, even this seemingly straightforward method can introduce a bias into the measurements, even if random statistical designs are in use. The reason for the bias is that the assistant reading the zone diameters sees the numerical measurement on the caliper's scales, then dictates these measurements aloud to the other assistant and hears the dictated figure through the ears. It can be seen, therefore, that the information is reaching the brain of the reader via three different channels, namely, sight, speech, and hearing. The net result is that after reading five or six zone diameters the reader's brain becomes attuned to the expected zone sizes, thus subconsciously achieving greater uniformity of replicate zone measurements than actually exists. This effect was demonstrated in the laboratory of one of the authors (Vincent) when the Autodata digital caliper was compared to the clock-dial manual calipers and to the Autodata automatic zone reader. Much to the surprise of the statisticians, zones measured with the clock-dial calipers achieved the lowest standard error. The manual measurements were expected to produce the highest error. In spite of the explanation of being influenced by several senses, as stated previously, the statisticians

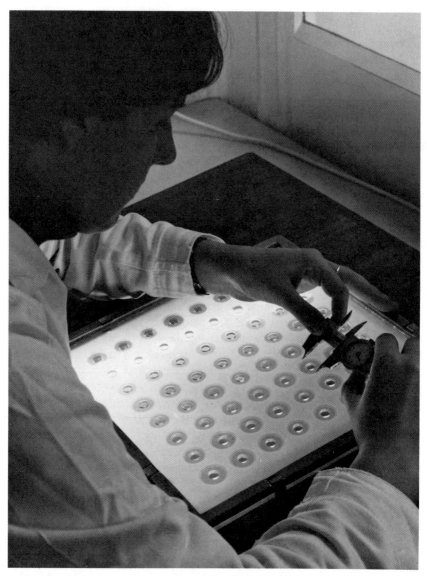

Fig. 4.10.   Using hand-held dial calipers for the measurement of zone diameters on a 30 × 30 cm large plate which is resting on top of an illuminated "reading box."

were not fully convinced. The comparison was repeated to eliminate a chance result, but the earlier observation was confirmed. In both of these tests the automatic zone reader gave the next lowest standard error and the digital calipers the highest. The highest error was the measure of the so-called biological error (i.e., the true human error), as opposed to the purely machine error produced by the automatic reader.

It is concluded that operator anticipation of zone size leads to a higher degree of reproducibility and falsely high apparent precision. These misleading results can be avoided by the use of thoughtfully and purposely designed equipment such as the Autodata digital caliper, which is described more fully later on.

Before describing other methods of making zone measurements, however, further practical hints are given here. Occasionally too much moisture collects on the surface of the agar during incubation. This happens more often if the assay medium was poured into the plates at a temperature of 60°C or higher and was not dried sufficiently before dispensing the test solutions. If this happens, lay a sheet of clean blotting paper on the agar surface and smooth it down gently so that it makes full contact with the agar.

Note: Place the wet blotting paper into a suitable "biohazard" bag for subsequent sterilization in an autoclave before discarding it.

On other occasions some or all the zones may be covered by surface growth which has to be removed before the zone edges can be seen clearly. The easiest way to do this is with paper tissue dipped into warm distilled water. Gently wipe the surface with the wet tissue in an up-and-down direction until all surface growth has been removed. Avoid left to right and right to left directions in case the zone diameters are affected.

Note: Wear disposable gloves for this operation, and discard the wet tissues and the gloves at the finish into biohazard bags for autoclaving.

Reading zone diameters is greatly facilitated by the use of specially constructed reading boxes. These are ventilated wooden boxes with a light source inside and glass mounted centrally on top. The glass is just large enough to accommodate one plate. The lid is suitably sloped to facilitate comfortable reading. The interior of the box is painted mat black under the glass, and indirect illumination is provided by strip lights on either side of the glazed center. A sketch of a suitable design is shown in Fig. 4.11.

The Autodata digital caliper has already been mentioned briefly for reading zone diameters. Figure 4.12 shows an early model in use. One of the advantages of this instrument is that the reading bias described earlier with the clock-dial calipers can be eliminated. This digital caliper has no visible

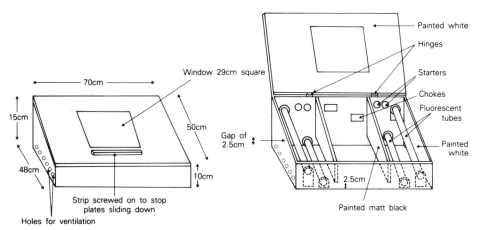

Fig. 4.11. Sketch showing the construction and dimensions of a typical illuminated reading box (not drawn to scale) to facilitate the measurement of zone diameters produced on 30 × 30 cm large plates.

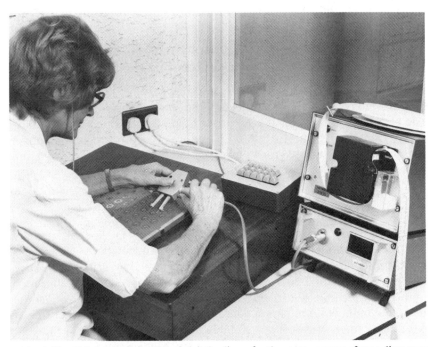

Fig. 4.12. Using an early model of digital calipers for the measurement of zone diameters on 30 × 30 cm large plates. The measurements are automatically transferred to paper tape. The small numerical keyboard is used for inserting header data, carriage return, and end marker.

scale; therefore, the measurement cannot be dictated, and because the zone diameter is not dictated, it cannot be heard. Upon determining that the needle points of the caliper are lined up with the opposing zone edges, the reader presses a button and the zone diameter reading appears on the electronic screen. At this stage the reading can be safely copied by another assistant or by the reader without in any way influencing the reader's judgment, because the button on the caliper will not function again until the needle setting has changed. A further advantage of this instrument is that it can be connected to a small printer or even to a desktop computer. A complete assay system including biocaliper, desktop computer, and associated printer, plus a large plate, is shown in Fig. 4.13.

Another convenient instrument for measuring zone diameters is the Fischer–Lilly reader, which has been available commercially since the 1950s. By aligning the projected zone edge to a marking line, one can read off the zone diameter to the nearest 0.1 mm with the aid of the calibrated dial which moves a tangential cursor. Wallhäußer (1982) describes a home-made reader which is equipped with data capture.

Autodata also markets a projector-assisted zone reader which produces an 8.75× magnified image of the zone. The instrument, connected to a

Fig. 4.13.   A current model of an Autodata complete assay system with hand-held digital calipers. Zone diameters are read directly into the computer's file, and results can be analyzed and printed after each plate is read or up to 40 sets of readings can be stored before printing. Two pages of results are printed in about 90 seconds.

desktop computer and printer, is illustrated in Fig. 4.14. A complete calculation package software is available with the instrument.

The problem with magnified images is that the projector also magnifies the imperfections of the zone edge, and if the zone definition is not sharp then the enlarged zone edges will become even more diffuse.

There are a few fully automated zone-reading instruments available from specialist firms, and these can completely eliminate human error. Some instruments measure zone areas from which the zone diameter or radius can be calculated. Most of these instruments use image analyzers coupled to computers and can be rather expensive. An early model of an Autodata automatic zone reader which has been in regular use for the past 15 years for measuring zone diameters of both antibiotics and vitamins on large plates can be seen in Fig. 4.15.

Fig. 4.14. A current model of an Autodata complete-assay system with projector. The left-hand edge of the enlarged zone is aligned to a fixed line on the screen and the cursor arm is lined up with the right-hand edge of the zone by means of a hand wheel. Pressing the center button of the hand wheel transfers the measurement of the zone diameter directly to the desktop computer. Programs are menu-driven, with prompts displayed at all stages to ensure correct operation.

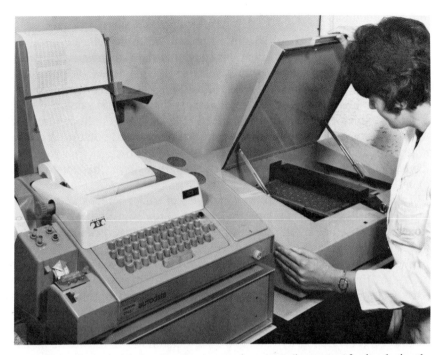

Fig. 4.15. Early model of an Autodata automatic zone-reading setup. After incubation the large plate is clamped into a mechanical carriage, which moves the plate to and fro from right to left and from left to right. Optical sensors measure the 64 zone diameters in less than 1 minute. The measurements are recorded by the printer and are transferred simultaneously to paper tape. Results and statistical analyses for each plate are calculated by a mainframe computer.

Fully automated instruments can be justified only in cases where a high throughput of samples is required and where good-quality zones can be maintained consistently. Instruments developed more recently, such as the ones shown in Fig. 4.13 and 4.14, can be purchased complete with mini-computer and user-friendly software which provide potency and confidence limits calculations. Desktop minicomputers have become relatively less expensive in recent years, and they are finding their way into microbiological laboratories for assay calculations and laboratory management. Large pharmaceutical companies have found it convenient in the past to purchase an instrument for simply recording zone data, which are then fed into the company's mainframe computer for a full statistical treatment.

4.6.6   Calculation of Potency Estimate

The nature of the dose–response line will be more fully discussed in Chapter 8. It will suffice to say at this stage that the dose–response line for the standard preparation is assumed to be straight when the measured response is plotted against the logarithm of the dose. Similarly, the dose–response line for a sample of the same active substance is also assumed to be straight. However, the actual potency of the sample is unknown and so responses must be plotted against a nominal or assumed potency. A consequence is that the two dose–response lines should be parallel; they would be coincident only if the assumed potency of the sample happened to be identical with its true potency.

In an assay comprising three dose levels of each standard and sample, means of responses to standard test solutions are designated S_1, S_2, and S_3 for low, medium, and high dose, respectively. Similarly, mean responses to the sample of unknown potency are designated T_1, T_2, and T_3.

Figure 4.16 illustrates an ideal plot of dose–response lines for such an assay. In this case a ratio of 2:1 between adjacent dose levels has been chosen; the actual potency of the sample is substantially higher than its nominal potency, and so its dose–response line is displaced upward from that of the standard. This graphic representation of the dose–response relationship is valuable in leading to an understanding of the mathematical expressions used in the calculation of potency estimate. First, the remainder of the symbols appearing in Fig. 4.16 must be defined: I is the logarithm (base 10) of the ratio of adjacent dose levels; in this case it is $\log 2 = 0.301$. M is the logarithm of the ratio of potencies of test solutions, sample : standard; this is the unknown factor to be determined in the assay. $\log 10$ corresponds to an extrapolated 10:1 dose ratio and has the value 1.000. E represents the mean difference in response arising from a dose increase of I on the log scale; i.e., it is the mean difference in response between adjacent dose levels. F represents the mean difference in response arising from an increase (or decrease) of dose by M on the log scale; i.e., it is the difference in response between the two preparations. b represents the theoretical (extrapolated) mean difference in response arising from a 10-fold increase in dose; i.e., it corresponds to an increase of $\log 10 = 1.000$ on the scale for logarithm of dose.

Figure 4.16 represents the ideal case when S_1, S_2, S_3, T_1, T_2, and T_3 lie exactly on two lines which are straight and parallel. In practice, of course, there will be deviations from this ideal. From the diagram it is shown that one estimate of the value E is given by:

$$E = S_2 - S_1$$

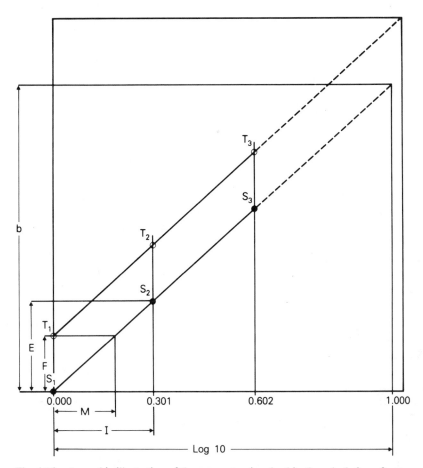

Fig. 4.16.   A graphic illustration of the parameters involved in the calculation of potency estimate in a three-dose level (3 + 3) parallel-line assay. See text for an explanation of symbols.

but equally valid estimates may also be obtained as $S_3 - S_2$, $T_2 - T_1$, and $T_3 - T_2$. Clearly, when there is random variation the best estimate of E is given by the mean of these four individual estimates as

$$E = \tfrac{1}{4}[(S_2 - S_1) + (S_3 - S_2) + (T_2 - T_1) + (T_3 - T_2)]$$

which simplifies to

$$E = \tfrac{1}{4}[(S_3 + T_3) - (S_1 + T_1)] \tag{4.1}$$

Similarly, from the diagram one estimate of F may be obtained as

$$F = T_1 - S_1$$

However, a better estimate is obtained as a mean of three individual estimates as

$$F = \tfrac{1}{3}[(T_3 + T_2 + T_1) - (S_3 + S_2 + S_1)] \qquad (4.2)$$

Again referring to Fig. 4.16, from a consideration of similar triangles it is clear that

$$E/I = F/M = b/\log 10 = b$$

A value for M may be calculated quite simply as

$$M = FI/E$$

without making use of the term $b/\log 10$. Potency estimate may then be obtained by taking the antilogarithm of M. However, for the statistical evaluation that is described in Chapter 8 it is necessary to obtain the value of b. Apart from this, b is a measure of the slope of the line and is valuable in the routine monitoring of the quality of the assay as described in Chapter 7. It is acknowledged that E also provides a measure of the slope of the line; however, it is dependent on the ratio between adjacent dose levels whereas b is independent of dose ratio. Thus, it is customary first to calculate

$$b = E/I \qquad (4.3)$$

then,

$$M = F/b \qquad (4.4)$$

Expressions for the calculation of values for E and F may be obtained by analogous derivation for any assay design having a constant ratio between adjacent dose levels. For convenience many of these in common or occasional use are presented in Appendix 6.

The calculation is illustrated in Example 4.1 by an assay of bacitracin in which the three dose levels of standard and sample are arranged, with a replication of six in a 6×6 Latin square design on a large plate. The Latin square design that was used in this assay is shown in Fig. 4.4.

Example 4.1: Assay of Bacitracin

Test organism: Micrococcus flavus
Design: 6×6 Latin square
Dose ratio: $2:1$
Weighings and dilutions for reference standard:
  Working standard potency: 55.3 IU/mg "as is"

$$95.7 \text{ mg} \rightarrow 100 \text{ ml}: 10 \text{ ml} \rightarrow 50 \text{ ml}: 10 \text{ ml} \rightarrow \ 50 \text{ ml high}$$

$$: 10 \text{ ml} \rightarrow 100 \text{ ml medium}$$

$$: 10 \text{ ml} \rightarrow 200 \text{ ml low}$$

Thus, the actual potency of the high-dose test solution is 2.117 IU/ml; for convenience it is regarded as having a nominal potency of 2 IU/ml with a factor of 1.0585 being applied.

Weighings and dilution for the "unknown" sample: The sample in this case was from a batch of zinc bacitracin and so was expected to have a potency about the same as the reference standard; thus, similar weighings and dilutions were made. The actual weight of sample taken was 96.3 mg.

Responses are shown in Table 4.1. From the means of responses to each treatment, values for E and F are calculated in accordance with Eqs. (4.1) and (4.2), respectively, thus:

$$E = \tfrac{1}{4}[(21.43 + 21.40) - (16.45 + 16.53)] = 2.4625$$
$$F = \tfrac{1}{3}[(21.40 + 19.20 + 16.53) - (21.43 + 19.17 + 16.45)] = 0.0267$$

then, using Eqs. (4.3) and (4.4),

$$b = 2.4625/0.30103 = 8.180 \text{ mm}$$
$$M = 0.0267/8.180 = 0.003264$$

from which potency ratio is obtained as antilog $M = 1.00754$. The potency of the high-dose "unknown" test solution is thus estimated to be

$$2.0 \times 1.00754 \times 1.0585 = 2.133 \text{ IU/ml}$$

from which the potency of the sample is estimated as

$$\frac{2.133 \times 50 \times 50 \times 100}{10 \times 10 \times 96.3} = 55.4 \text{ IU/mg}$$

Table 4.1

Responses in the Assay of Bacitracin (Example 4.1)

	Zone diameters (mm)					
	Sample			Reference standard		
Dose level:	High	Medium	Low	High	Medium	Low
	21.9	20.0	16.6	21.9	19.9	16.8
	21.7	19.7	16.7	22.1	19.4	16.7
	21.8	19.6	17.1	22.0	19.4	17.0
	21.1	19.1	16.7	21.2	19.0	16.5
	21.6	19.1	16.5	21.0	19.0	16.1
	20.3	18.0	15.6	20.4	18.3	15.6
Totals:	128.4	115.5	99.2	128.6	115.0	98.7
Means:	21.40	19.20	16.53	21.43	19.17	16.45

4.6.7 Safe Disposal of Contaminated Material

The majority of microorganisms used in microbiological assays are harmless, but because of the astronomical numbers involved, they need to be treated with respect. Under current good laboratory practice, all workers are responsible not only for their own safety but also for the safety of their fellow workers. For this reason any plastic or glassware which was in contact with live microorganisms must be carefully segregated and rendered harmless before being washed.

In contrast, volumetric glassware used for standard and sample solutions can be safely washed without the need for decontamination.

Having measured the zone diameters, transfer all the disposable petri dishes containing inoculated assay medium into autoclavable biohazard bags and, without tying or sealing the bags, put them into suitable aluminum or stainless-steel decontamination bins which are supplied with loose-fitting lids. Ideally, load the full or partially filled bins into a double-ended autoclave on the "dirty side," and sterilize the contents by using a suitable decontamination cycle (e.g., $126\,°C$ for 1 hour). After sterilization, remove the bins when they are safe to handle on the "clean side" of the autoclave. Pack the biohazard bags, together with their contents, securely into stiff cardboard boxes, and have them incinerated as soon as possible. Bear in mind that the previously used medium which has now been sterilized is still capable of supporting microbial growth and could become heavily contaminated within a few days. However, residues of media should not pose any greater public hazard than, say, kitchen waste.

Bottles, test tubes, flasks, or glass petri dishes, and similar items containing residues of seeded agar, out-of-date slant cultures, or bacterial suspensions, must be put through a decontamination cycle just like the disposable items, before they can be washed. It is convenient to pack contaminated glass bottles, glass petri dishes, and the like, into suitable wire baskets for easy handling. Transfer the filled baskets into the bins used for decontamination, as described earlier. The bins can be made locally in any convenient size to fit the available autoclave. A suitable decontamination bin is shown in Fig. 4.17.

Attach a tie-on reusable metal or other heat-resistant label with "BIO-HAZARD" printed or engraved on it as soon as any contaminated material has been put into the bin. Only authorized and bacteriologically trained staff should be allowed to take off the lid of these bins before decontamination, and then only for the purpose of putting more contaminated items into them.

There may be occasions when it is necessary to wash large plates without autoclaving when time is short and the plates are urgently needed or when

Fig. 4.17. A typical decontamination bin. Any glassware and plastic disposable items that were in contact with live bacteria, yeasts, or molds are deposited in these bins for sterilization by autoclaving before either washing or safe disposal. The bin itself is leakproof, but the loose-fitting lid is provided with perforations.

the autoclave is out of action. Large plates of the Autodata demountable design can be treated safely as follows:

Wearing disposable gloves, take off the lid and immerse it in a suitable disinfectant solution, such as a 2% (v/v) aqueous solution of Hycolin. Remove the spring clips with the aid of a suitable tool such as a strong metal spatula, and drop them into a small jar or beaker containing disinfectant solution. Carefully lift off the metal frame without disturbing the agar on the glass plate, and immerse it in the disinfectant solution containing the lid. Hold the glass plate with the agar layer on top over a stainless-steel bucket or other suitable container, and peel off the agar layer with the aid of a wide, flat spatula so that the agar drops into the bucket. Now immerse the stripped glass plate also into the disinfectant solution, which already contains the lid and frame. Continue stripping the plates in this manner until all the plates have been dealt with; then cover the bucket with a loose-fitting lid and sterilize it in the autoclave at the first opportunity.

After autoclaving, the hot, molten agar either can be washed down the drain with plenty of hot water or can be allowed to solidify, after which it can be transferred into biohazard bags for incineration. The lids, frames, clips, and glass plates can be safely washed after about 30 minutes' contact with the disinfectant solution. Pay particular attention to the corners and sides of the frames by scrubbing with a nailbrush in soapy water, followed by several rinses of tap water and a final rinse of distilled water. Regular use of soap-filled steel-wool pads (e.g., Brillo pads) on the frames and lids will keep them sparkling clean.

From time to time it is advisable to soak all glassware in a suitable detergent (e.g., a 1% (w/v) solution of Pyroneg or dilute Decon solution) to eliminate all traces of any active ingredients that are known to cling to glass surfaces. Lightbown (1975) reported the development of "ghost zones" on large plates due to inadequate cleaning procedures. Refer to Section 7.14 for further details. Another method that can be tried to eliminate really tenacious residues like neomycin and nicotinamide is to adopt the following heat treatment as needed:

Stack the loosely assembled large plates (without clips) into a hot-air oven when the oven is cold, switch on and allow the temperature to reach about 160°C; then switch off and let the plates cool naturally to room temperature. Store the heat-treated plates in a dust-free area until required. Wash the clips separately and, after drying, store them in a suitable box fitted with a lid.

Large plates which have had the sides glued on need extra care, because the glued-on glass strips can be partially lifted, allowing either disinfectant or detergent solution to seep under the strip; this can cause inhibition around the edges after incubation of the seeded assay medium. Should this happen, it is better to remove the strip altogether, clean off all traces of the old adhesive, and reglue the cleaned glass strips.

Glass-edged large plates do not stand up to the rigors of autoclaving or to the heat processing described previously because of the problem of loosening glass strips. The best cleaning routine to adopt with these plates is simply to strip the agar layer into a stainless-steel bucket with lid and decontaminate the medium by autoclaving. Soak the stripped plates and the lids in a suitable disinfectant solution for about 30 minutes, rinse under the tap, and immerse them in Pyroneg or Decon-90 solution (or some other suitable detergent) for at least 20 minutes. Finally, wash in the sink with plenty of tap water and dry with a clean cloth.

Cleaning of agar cutters is described in Section 7.11.3 and the safe handling of contaminated graduated pipettes in Section 2.2.

References

British Pharmacopoeia (1980). British Pharmacopoeia Commission, H.M. Stationery Office, London.

Cooper, K. E. (1963 and 1972). *In* "Analytical Microbiology" (F. W. Kavanagh, ed.), Vols. I and II. Academic Press, New York and London.

Cooper, K. E., and Gillespie, W. A. (1952). *J. Gen. Microbiol.* **7**, 1.

Cooper, K. E., and Linton, A. H. (1952). *J. Gen. Microbiol.* **7**, 8.

Cooper, K. E., and Woodman, D. (1946). *J. Pathol. Bacteriol.* **58**, 75.

European Pharmacopoeia (1980). European Pharmacopoeia Commission, Strasbourg.

Hans, R., Galbraith, M., and Alegnani, W. C. (1963). *In* "Analytical Microbiology" (F. W. Kavanagh, ed.). Academic Press, New York and London.

Hewitt, W. (1977). "Microbiological Assay." Academic Press, New York and London.

Humphrey, J. H., and Lightbown, J. W. (1952). *J. Gen. Microbiol.* **7**, 129.

Lees, K. A., and Tootill, J. P. R. (1955a). *Analyst* **80**, 95.

Lees, K. A., and Tootill, J. P. R. (1955b). *Analyst* **80**, 531.

Lightbown, J. W. (1975). Personal communication.

Mitchison, D. A., and Spicer, C. C. (1949). *J. Gen. Microbiol.* **3**, 184.

Simpson, J. S. (1963). *In* "Analytical Microbiology" (F. W. Kavanagh, ed.), Vol. I. Academic Press, New York and London.

Vincent, S. (1970). Unpublished information.

Wallhaüßer, K. H. (1982). *Pharm. Ind.* **44**, 301.

CHAPTER 5

TUBE ASSAYS FOR GROWTH-INHIBITING SUBSTANCES

5.1 Introduction

For convenience and brevity, the phrase tube assay will be applied only to the assay methods using liquid medium and not to the tube version of the agar diffusion assay that was mentioned in Chapter 4. The general pattern of growth of a microorganism in a liquid nutrient medium which will be described is of relevance both to the assay of growth-inhibiting substances (the topic of this chapter) and to the assay of growth-promoting substances, the topic of Chapter 6. When a small inoculum of a microorganism is added to a liquid medium containing all the nutrients needed for its growth and then that inoculated medium is kept at a suitable temperature, growth proceeds in accordance with the general pattern shown in Fig. 5.1. After a short period of slow growth, the lag phase, growth proceeds at a steady rate on a logarithmic scale. This is followed by a period when the population changes little, due to depletion of the nutrient medium and possibly also due to the buildup of waste materials having a bacteriostatic effect. Eventually the concentration of living cells begins to decline as medium becomes more depleted, and concentration of bacteriostatic substances may become even greater. These four phases are known respectively as (i) the lag phase, (ii) the log phase, (iii) the stationary phase, and (iv) the decline phase.

Provided that the medium includes adequate amounts of all the nutrients, no growth-inhibiting substances are present, and the temperature for growth is optimal, then the lag phase will be short. On passing to the log phase of growth, the population of living cells doubles in a period of time known as the generation time. The generation time is dependent on the innate characteristics of the organism, the nature of the medium, and the temperature, but may also be modified by the presence of growth-inhibiting substances. Generation times may range from about 15 to over 120 minutes for bacteria, from about 40 to 180 minutes for yeasts, and from about 90 to 360 minutes for molds.

When growth is allowed to proceed as far as the stationary phase, the log phase might, typically, last for about 15 generations. This, naturally, is dependent on the size of the inoculum. An inoculum of 10,000 viable cells/ml would increase in 15 generations approximately thus:

$$10,000 \times 2^{15} = 3.28 \times 10^8$$

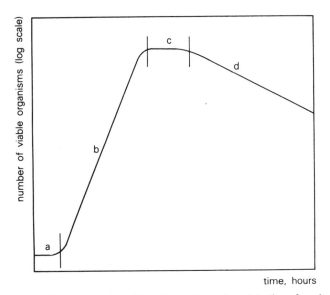

Fig. 5.1. A graphic representation of the phases of growth and decline of a culture. a, Lag phase; b, log phase; c, stationary phase; d, decline phase.

Further growth during the transition from log phase to stationary phase might bring the total cell count up to 10^9/ml.

It was shown by Brown and Garrett (1964) that for one strain of *Escherichia coli,* generation time was constant (~25 minutes) regardless of inoculum level. Inoculum levels ranging from 10^3 to 10^7 cells/ml were used. The log phase ended when a viable cell concentration of about 4×10^8/ml was reached. The maximum level of viable cells attained at the high point of the stationary phase was about 8×10^8/ml.

The apparently very large number of cells that are eventually produced still represents only a very dilute suspension. If, for example, there were 10^9 cells/ml, they were spherical, and they had a diameter of $1\,\mu m$, the total volume occupied by them would be only 0.000524 ml, or about 0.05% of the total volume of the liquid suspension.

5.2 Outline of the Method

The tube assay for growth-inhibiting substances is simple in principle. Test solutions at a series of concentrations are prepared from both reference standard and sample(s) to be tested. In a typical assay, 1 ml of each test solution is added to each of three individual test tubes. To three more tubes 1 ml of water is added to serve as zero-dose controls. Chilled nutrient broth is inoculated uniformly with a test organism sensitive to the growth-inhibiting substance to be assayed, then 9 ml of the inoculated broth are added to each tube. The tubes, in racks, are then placed in a well-stirred water bath at the required incubation temperature, where they remain for about 3–4 hours. When turbidity in the zero-dose control tubes is sufficient to give an adequate optical response, growth is terminated in all tubes at the same time. Turbidity, as an indication of total growth, is then measured in all tubes using a nephelometer or an absorptiometer.

Total growth of the organism is reduced to an extent dependent on the concentration of the growth-inhibiting substance. A comparison between the reductions in growth in standard and sample series of tubes is the basis of the estimation of potency.

5.3 Nature of the Response

Whichever method is used to measure the optical response, it is for some purposes convenient to express response (or mean response) to an individual test solution or "treatment" as "proportionate response," which is defined as

$$\frac{\text{mean response to treatment}}{\text{mean response to zero dose}}$$

A plot of proportionate response versus logarithm of dose over a sufficiently wide range of doses is sigmoid, as illustrated qualitatively in Fig. 5.2. The ranges a–b and e–f correspond to uninhibited growth and to complete inhibition, respectively. The central region c–d is almost straight and is potentially the most useful range for assay purposes. However, the slope of this region may be so steep that it corresponds to only a very narrow range of dose levels. In routine assays the complete sigmoid curve is not generally displayed.

The possible mechanisms by which a growth-inhibiting substance causes a reduction in the final cell population in an assay tube as compared with that of another assay tube having zero dose of growth-inhibiting substance include (i) an increase in the duration of the lag phase, (ii) an increase in the generation time, and (iii) killing of a proportion of the cells. A combination of two or more of these postulated mechanisms is also possible.

Garrett and Miller (1965) studied the effect of different concentrations of tetracycline on the growth of *E. coli* in the log phase over a period of 6 hours and compared the changes in total number of organisms (by Coulter counter) and number of viable organisms (by plate count). It was found that at concentrations from about 50 to 200 ng/ml of tetracycline, total and

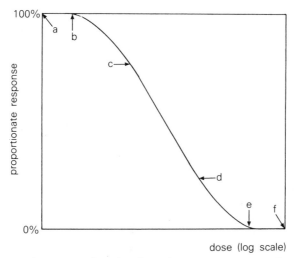

Fig. 5.2. A graphic representation of the form of a dose–response curve in the turbidimetric assay of a growth-inhibiting substance. Dose is shown on a logarithmic scale and response (cell concentration) on an arithmetic scale. Ranges of curve shown: a–b, uninhibited growth; e–f, complete inhibition; c–d, central region, which may approximate to a straight line and is the most useful range for assay purposes. Normally, the range of doses used in a routine assay would not be sufficiently wide to show the whole range of responses depicted here.

viable counts coincided closely, and that the rate of growth decreased as tetracycline concentration increased. However, at about 4 μg/ml of tetracycline, the total cell count remained unchanged while the viable count decreased with time. These observations are illustrated graphically in Fig. 5.3. They suggest that at concentrations of up to about 200 ng/ml tetracycline acts only by increasing the generation time, but at a concentration of about 4 μg/ml multiplication of the organism ceases and some cells begin to die. The inference that at lower concentrations of tetracycline the only mode of action is to increase generation time is based on the presumption that a killed organism will cause a response in the total count method. This is not necessarily so if the killing of the cell is through lysis. Observations similar to these have been obtained in the cases of some sulfonamides, lincomycin, and erythromycin.

Equations representing mechanisms (i) and (ii) either singly or combined were derived by Kavanagh (1968) and further developed by him (1975). Kavanagh concluded that in most of the cases of interaction between a growth-inhibiting substance and a growing organism that had been studied, the main effect was in accordance with mechanism (ii). In such cases the equations indicated that a plot of logarithm of final cell count against

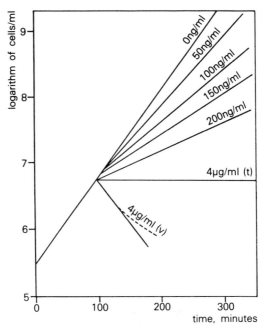

Fig. 5.3. The influence of varying concentrations of tetracycline on the growth of *Escherichia coli*. At tetracycline concentrations of 50–200 ng/ml (1–4×10^{-7} M), total and viable counts were almost identical and are represented by a single line for each concentration. However, at 4 μg/ml (8×10^{-6} M), viable counts decreased with time (v), whereas total counts remained constant (t).

concentration of growth-inhibiting substance should be a straight line. In practice, of course, it would be more convenient to plot logarithm of some optical measure of cell concentration against the concentration of growth-inhibiting substance. An example of such a plot is given in Fig. 5.4, which is the standard curve from an assay of erythromycin using *Staphylococcus aureus* as test organism. Here, the logarithm of optical absorbance at 530 nm is plotted against concentration of antibiotic. It will be noted that from 2 to 5 μg/ml the plot is an almost perfect straight line. However, the response to zero dose does not lie on the straight line. Samples were assayed at only one dose level each and so no sample dose–response lines can be shown. It may be noted that if samples had been assayed at more than one dose level, their dose–response lines should also be straight in the same dose range and should intersect with that of the extrapolated standard

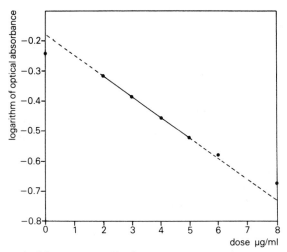

Fig. 5.4.   A standard dose–response line for erythromycin in the tube assay using *Staphylococcus aureus* as test organism. The plot of logarithm of optical absorbance against concentration of erythromycin is linear in the dose range 2–5 μg/ml.

straight line at its point of intersection with the *y* axis; i.e., this would be a "slope ratio assay" (see also Sections 6.5.5 and 8.3.1). The raw data from which Fig. 5.4 was derived are presented in Table 5.1.

Table 5.1

Raw Data for the Standard Dose–Response Line in the Turbidimetric Assay of Erythromycin Using *Staphylococcus aureus* as Test Organism[a]

Dose (μg/ml)	Observed absorbance at 530 nm		Mean absorbance	Logarithm of mean absorbance
0	0.569	0.575	0.572	−0.243
2	0.484	0.480	0.482	−0.317
3	0.414	0.411	0.4125	−0.385
4	0.350	0.350	0.350	−0.456
5	0.300	0.300	0.300	−0.523
6	0.265	0.264	0.2645	−0.578
8	0.213	0.212	0.2125	−0.673

[a] From the work of Frederick Kavanagh. The assay was done using the Autoturb system; both 0.1-ml and 0.15-ml sampling loops were used in the assay. The data here show only the observations obtained when using the 0.1-ml loop.

If, instead of mechanisms (i) and (ii), reduction in final cell concentration were due to mechanism (iii), then it would be reasonable to suppose that the varying sensitivity of individual cells to a growth-inhibiting substance, if plotted against logarithm of dose, would form a "normal distribution." The normal distribution curve, which is of fundamental importance in the science of statistics, is shown in Fig. 5.5a, together with the related cumulative distribution curve of Fig. 5.5b.

Procedures for the linearization of sigmoid response curves based on the relationships illustrated in Fig. 5.5a and b were first proposed independently by Hemmingson (1933) and Gaddum (1933). These procedures were devised for use in assays where the response is quantal, i.e., of the "all-or-none effect" type such as the death of a test animal. Percentages of deaths in a group of animals (the proportionate response) were related through the cumulative distribution curve to the "normal equivalent deviation," or, as it is more commonly termed now, the "standard deviation."

The graphs in Fig. 5.5a and b, which are of general application in scientific studies, are used in the present context to represent the frequency distribution of responses. Thus, the area under the normal distribution curve lying between ± 1 normal equivalent deviations is approximately 67% of the whole area and corresponds to the same percentage of responses. Similarly, the area between ± 2 normal equivalent deviations corresponds to about 95% of all responses. The term probit was introduced by Bliss (1934a,b) as an abbreviation of "probability unit," which is derived from the normal equivalent deviation simply by adding 5 to it to eliminate negative figures and to facilitate calculation procedures.

The relationship between percentage response, normal equivalent deviation, and probit is shown in Table 5.2. Extended tables relating percentage response and probit are given by Fisher and Yates (1963). Although there

Table 5.2

Relationship between Percentage Response,
Normal Equivalent Deviation, and Probit

Percentage response	Normal equivalent deviation	Probit
2.28	−2.0	3.0
15.87	−1.0	4.0
50.00	0.0	5.0
84.13	+1.0	6.0
97.72	+2.0	7.0

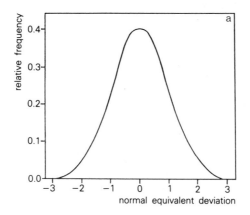

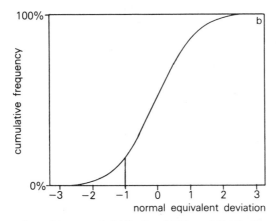

Fig. 5.5. Illustrations of the normal distribution. (a) Normal distribution curve. Its vertical axis indicates the relative frequency of observed values in a normally distributed population according to their deviation from the mean. The total area under the curve is unity; the areas between ± 1 and ± 2 normal equivalent deviations are 0.683 and 0.955, respectively, of the whole area. (b) Cumulative frequency curve. It is related to (a) and shows the area under the curve in (a) lying to the left of any vertical line drawn to designate the normal equivalent deviation. Thus, as in curve (a), the area lying under the curve between ± 1 normal equivalent deviations is 0.683; the area under the curve but outside these limits must be $1.000 - 0.683 = 0.317$; and the area to the left of the line representing -1 normal equivalent deviation must be $0.317/2 = 0.1585$, or 15.85% as shown in (b).

appears to be little evidence of the existence of mechanism (iii), the sigmoid curve of Fig. 5.2 does resemble the curves obtained in quantal assays and the probit transformation is, in practice, found to be useful.

5.3.1 The Response in Practice

Whatever the mechanism of action of a growth-inhibiting substance, the net result in the cases of many assays is that over a sufficiently wide range of doses, a series of log dose–response lines such as are illustrated in Fig. 5.6 is produced. The range of doses, however, is often insufficiently wide to yield the complete sigmoid curve. If, as shown in Fig. 5.6, the central regions of both standard and sample response lines are almost straight and parallel, then an assay design and calculation procedure analogous to those shown for the agar diffusion assay in Chapter 4 may be applied. The characteristics of the central region of the response lines may be defined by two parameters: the median response and the slope. The median response *(MR)* is the dose of growth-inhibiting substance that permits growth of the organism to 50% of the level attained in the zero-dose control tube in the same assay system. It is a measure of the activity of the growth-inhibiting substance under the specific assay conditions.

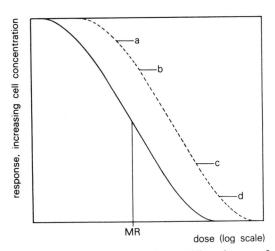

Fig. 5.6. A theoretical illustration of responses in a turbidimetric assay of a growth-inhibiting substance. This shows the case when the actual potency of the sample is substantially less than its nominal potency, hence the curve for sample responses is displaced to the right of that for standard. The central region of responses (between points b and c) approximate closely to straight and parallel lines and so, clearly, meet the criteria for a parallel-line assay. However, experience has shown that responses in the range a–d could probably be used for assay purposes, applying the standard calculation procedures, with little error despite the curvature. *MR*, Median response to reference standard.

The slope b of the response line is expressed as the change in the cell concentration (or perhaps turbidity) of the cell suspension corresponding to a 10-fold change in dose of growth-inhibiting substance; it has a negative value in these assays. (The value b is necessarily calculated from a change in dose level much less than 10-fold; it expresses the rate of change in response with changing dose.) The possible variation in slope of this central region is illustrated in Fig. 5.7, which is based on the data of Rippere (1979). The slopes were monensin, -0.18; gramicidin, -0.75; and erythromycin, -2.20. Rippere considers the slope for gramicidin to be the most satisfactory of the three. The slope for monensin is inadequate, whereas a slope as steep as that for erythromycin is troublesome in that replicate tubes show a wide variation in the amount of growth and there is substantial daily variation in the useable portion of the response line.

The sensitivity of an assay system may be varied by changing the pH. As a general rule in the case of basic antibiotics such as streptomycin and erythromycin, an increase in pH leads to an increase in sensitivity (decrease in MR), but in the case of acidic antibiotics such as the penicillins and cephalosporins it leads to a decrease in sensitivity (increase in MR). In the assay of neutral antibiotics such as chloramphenicol, pH is generally without influence on sensitivity.

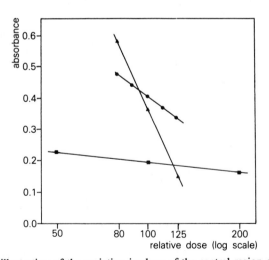

Fig. 5.7.   An illustration of the variation in slope of the central region of the log dose–response lines in turbidimetric assays of some growth-inhibiting substances. The substances, test organisms, and calculated slopes were as follows: monensin (■), *Streptococcus faecium* ATCC 8043, $b = -0.18$; gramicidin (●), *S. faecium* ATCC 10541, $b = -0.75$; erythromycin (▲), *S. faecium* ATCC 10541, $b = -2.20$; From the work of Rippere (1979).

Typically, an increase in inoculum increases MR and reduces the slope of the response line.

Attention is drawn to possible pitfalls in changing the pH of an assay medium by Kavanagh (1972). If the antibiotic is not a single active substance, then its different components may not respond similarly to a change in pH. It follows that if the standard and sample contain different proportions of the two or more active components, then a change in pH would lead to a change in estimated potency. This, of course, is one of the inevitable hazards of assaying mixtures of active substances. Of the differing estimates that would be obtained for different pH values, none could be said to be more correct than any other.

A change in pH does not generally influence the slope of the dose–response line. Although variations in slope have been demonstrated, slopes in this potentially useful part of the dose–response line are generally steep, with the consequence that the possible working range of dose levels is narrow. For this reason, close dose ratios between adjacent levels are necessary. Dose ratios such as $2^{0.5}:1$, $5:4$, $6:5$, and $10:9$ approximately are suggested by the United States Code of Federal Regulations (CFR).

It is of interest to note the dose ratios used in the assays reported by Rippere (1979). For the assay of gramicidin, five dose levels spanning a range from 80 to 125% from the midpoint correspond to a ratio between adjacent doses of $10:9$. For erythromycin spanning the same range with only three dose levels, the ratio between adjacent doses was $5:4$.

Parallel-line designs as applied to tube assays for growth-inhibiting substances must thus differ from those applied to the agar diffusion assay in that (i) narrower dose ratios must generally be used; and (ii) because of the substantial daily variation in the useable portion of the dose–response line, a greater number of dose levels must be set up to ensure that the useable range is covered. Responses to some dose levels may lie outside the useful working range and therefore have to be excluded from the calculation of potency estimate. It is, of course, possible to estimate the potency of the sample by interpolation from the standard log dose–response line even though it is curved. This, however, has the inevitable disadvantage of the errors inherent in judging by eye the curve of best fit. A purely mathematical estimation of potency is preferable whenever possible, so as to avoid this source of error. Having recommended mathematical methods, attention must be drawn to the possible pitfalls of their application in inappropriate circumstances. Hewitt (1987) observed that in a laboratory in which potency estimates were calculated by a computer, each estimate was 4.6–4.8% lower than the best estimate that could be made from the observations! This was due to the application of the official U.S. CFR calculation to a line that was distinctly curved. This constant error would have been readily brought to light by drawing the curve and interpolating to obtain a potency estimate.

Proper mathematical estimations depend ultimately on the existence of, or the derivation of some form of linear relationship between dose or function of dose and response or function of response. The relationship between the logarithm of the dose and probit of response has been mentioned in Section 5.3. This would, ideally, convert the sigmoid curve into a straight line. In practice it may be only partially successful but will probably at least provide a straight-line response for all but the extreme doses. Thus it will serve a very useful purpose in facilitating an arithmetic processing of the experimental data. Other transformations of response that may be used are the logit and the angular transformation. These are applied empirically and generally achieve the same result as the probit transformation. In fact for responses between 20 and 80%, there is virtually no difference in the effect of the three functions. The two newly introduced functions are defined thus:

Angular transformation:

$$p = \sin^2 \phi \qquad\qquad (5.1)$$

where p is the proportionate response and ϕ is an angle between $0°$ and $90°$.

Logit transformation: The logit, z, is defined by Fisher and Yates (1963) as

$$z = 0.5 \ln(p/q) \qquad\qquad (5.2)$$

where p is the proportionate response and

$$q = (1 - p) \qquad\qquad (5.3)$$

These transformations too generally result in a log dose–function of response relationship in which all but those responses to doses at either extreme of the range lie on or close to a straight line. Such extreme responses may be discarded and then the potency estimated on the basis of, preferably, a symmetric assay design calculation such as a $4 + 4$, $5 + 5$, or $6 + 6$ design. However, asymmetric designs such as a $4 + 3$ or a $5 + 4$ could be used depending on the numbers of responses lying on the straight part of the modified response lines for standard and "unknown."

5.4 Critical Factors

In the turbidimetric tube assay for growth-inhibiting substances, the measured response is dependent on the number of organisms in the individual tubes when growth has been terminated. It is pertinent to inquire, therefore, how many factors other than concentration of growth-inhibiting substance may influence this number. Time and temperature of incubation are clearly factors to be considered. In the growth-inhibiting substance tube assay, a level of turbidity convenient for measuring optically is generally attained in 3–4 hours—a time when growth is normally well within the log phase. The need to control strictly the time and temperature of incubation may be seen from the following theoretical considerations:

Suppose that at 35°C the generation time of the organism in the zero-dose control tubes is 20 minutes. If the duration of incubation in the log phase were exactly 4 hours, then there would be 12 generations during this period and cell concentration would increase by a factor of $2^{12} = 4096$. Consider now growth in the tube having a sufficient concentration of growth-inhibiting substance to lead to the median response; i.e., cell population would have increased by a factor of $4096/2 = 2048$ during the 4-hour period. Assuming that the growth-inhibiting substance has exerted its influence only through increasing the generation time, as the evidence of Garrett and Miller (1965) suggests, then in the tube showing median response there will have been 11 generations during the incubation period ($2^{11} = 2048$), and so generation time must have increased to $240/11 = 21.8182$ minutes. Suppose now that, due to the mechanics of handling a large number of tubes, another tube containing exactly the same concentration of growth-inhibiting substance is incubated for 241 minutes, then this period corresponds to $241/21.8182 = 11.046$ generations, and so cell population will increase by a factor of $2^{11.046} = 2114$. This latter figure (2114) is 3.23% greater than 2048. Thus the need for strict control of incubation time is clear.

Suppose now that in all tubes the incubation period were kept to exactly 240 minutes during the log phase of growth but that due to uneven temperature of incubation the generation time in a tube containing the same dose of growth-inhibiting substance as in the previous example were increased from 21.8182 minutes to 22 minutes (an increase of only 0.83%); then during the 240 minutes incubation there would be $240/22 = 10.909$ generations, and so cell concentration would increase by a factor of $2^{10.909} = 1923$. This figure (1923) is only 93.9% of 2048. One example of the effect of incubation temperature on generation time is given in Fig. 5.8, which is based on the work of Cooper (1963). This refers to the growth of *Staphylococcus* Mayo in liquid medium and shows that generation time changes steeply at temperatures not close to the optimum.

It is clear from these examples that control of both temperature and duration of incubation is of utmost importance. Incubation temperature should be close to optimum to minimize temperature-related changes in generation time.

It is also clear that any variation in the size of the inoculum from tube to tube will be reflected in the final cell populations of individual tubes.

5.5   The Tube Method in Practice

The unit operations of the tube assay method for growth-inhibiting substances are shown as a flow diagram in Fig. 5.9 and are numbered 1 – 12 for convenience. Steps 2 and 3 (maintenance of stock cultures and preparation of the inoculum) are described in Chapter 2; these must be done in advance of the day of the assay. It is also generally more convenient to prepare sterile nutrient medium (step 1) before the day of the assay and keep it refrigerated, thus avoiding the need to sterilize then chill immediately before use in the assay. Stock solutions of reference standards might be prepared in advance depending on their stability; however, preparation of the final dilute test solutions (step 5) must be done on the day of the assay so as to minimize loss of potency by decomposition. Because incubation (step 8) is normally of only about 4 hours duration, all other steps between 4 and 10 must be completed in one day. Good laboratory practice demands that step 12, disposal of used media and cleaning of glassware, also be carried out

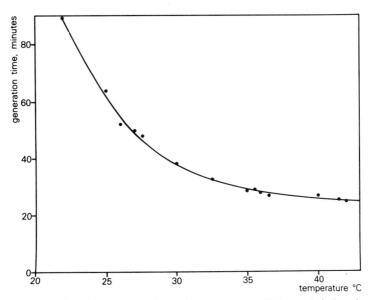

Fig. 5.8.   Variation of generation time with temperature. This graph is based on the observations of Cooper (1963) and refers to the growth of *Staphylococcus* Mayo in a beef heart infusion.

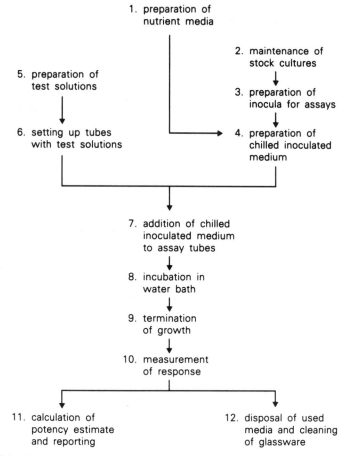

Fig. 5.9. Flow diagram showing unit operations in the tube assay for growth-inhibiting substances.

the same day. It may also be required that potency estimates be calculated and reported (step 11) the same day.

Thus, it is clear that for successful assaying by this method, good organization of the work is essential.

5.5.1 Setting Up the Assay Tubes

The three initial pathways shown in Fig. 5.9 finally unite at step 7: the setting up of the assay tubes containing the chilled, inoculated nutrient broth together with one of the test solutions or an equivalent volume of water in the case of the zero-dose controls.

Tubes of 12 mm diameter are convenient. To maintain a uniform rate of heating, the tubes must be of uniform dimensions and uniform thickness of glass. In preparation for the assay they must be provided with suitable caps, e.g., of aluminum. Both tubes and caps must be sterile. Each tube must be marked near its top with the designation of the test solution that it is to contain, using a suitable marking pen.

Precautions must be taken to avoid the ingress of extraneous organisms, because contamination by an organism not sensitive to the growth-inhibiting substance could be disastrous. Thus, in an unfavorable general environment, work in a laminar-flow cabinet might be desirable.

As was indicated in Section 5.2, in a typical assay 1 ml of test solution (or 1 ml of water) is added to each tube. This is then followed by 9 ml of the chilled inoculated broth. Bearing in mind the critical factors described in Section 5.4, it is clear that volumes of test solutions must be measured as accurately as practicable; an error in measurement of volume of test solution would not only affect the concentration of growth-inhibiting substance, but would also affect the concentration of the nutrient medium, albeit slightly, which could lead to a significant change in generation time. The potential effect of even a slight change in generation time has already been illustrated. Test solutions should therefore be measured using a 1-ml bulb pipette calibrated "to deliver." Class A pipettes are desirable for this purpose. In adding the 9 ml of chilled, inoculated broth, although the absolute volume is not critical, it is essential that variation in volume from tube to tube be minimal. Precautions must also be taken to avoid the bias that may arise from difference in time of addition of the broth to the tubes containing the test solutions. It is also important that the test solution and inoculated broth be properly mixed. The tubes, which have hitherto been arranged in an orderly manner in racks for convenience of adding test solutions, may now be rearranged in racks in accordance with a randomized distribution pattern. Suitable designs are shown in Appendix 8. This randomization tends to eliminate any bias that might have arisen from differences in time of addition of the nutrient medium, which could result in different times of warm-up to incubation temperature and so, effectively, different incubation periods. The addition of the chilled inoculated broth may be achieved conveniently by means of a high-speed semiautomatic pipetting device. It is necessary that any such pipetting device must be demonstrably capable of delivering a constant volume with minimum error. A high speed of addition is desirable for two reasons: differences in warm-up time are minimized, and the fast-flowing jet of liquid medium can help to ensure—through good mixing with the test solution—uniform distribution of the growth-inhibiting substance throughout the content of each tube.

5.5.2   Incubation and Termination of Growth

It was shown in Section 5.4 that time and temperature of growth are highly important factors in determining the final number of cells in each tube. Ideally then, all tubes should have the same temperature–time profile during warm-up to incubation temperature and should remain at that temperature for exactly the same period before growth is terminated. During the setting up of the tubes, steps are taken to facilitate attainment of the desired conditions for incubation. Thus, tubes of uniform shape and thickness of glass are used. The nutrient medium is chilled before being inoculated with the test organism and before being added to the tubes, so as to avoid commencement of growth before tubes are immersed in the incubator bath.

Tubes in racks are necessarily all placed in the incubator bath at the same instant. An incubator bath and not an incubator cabinet is used so as to ensure rapid and uniform warm-up of the tubes. The bath must be of such design as to ensure uniform temperature throughout; to achieve this it must, of course, be very well stirred. The random distribution of the tubes in racks would balance out any minor variations in temperature that exist despite the use of a well-designed water bath.

Growth must be terminated at the same instant in each tube in an assay. Kavanagh (1972) recommends immersion of the racks of tubes in a water bath at 85°C. An alternative preferred by some analysts is to use a high-speed dispenser to squirt 1 ml of formaldehyde solution into each tube. In this procedure, speed in completing the addition of formaldehyde to all tubes in an assay is essential. The randomization of the tubes helps to balance out the effective differences in incubation time that arise inevitably in this procedure. The squirting of the formaldehyde is intended to ensure good mixing and thus termination of growth immediately after the addition.

5.5.3   Measurement of Response

To this point in the chapter, response has been considered in terms of reduction in the number of cells due to the presence of a growth-inhibiting substance as compared with the number of cells in the zero-dose control tubes.

Much has been written about the variation in size of cell with the age of a culture and its consequent effect on the optical properties of the cell suspension. It is recognized that the optical properties of the cell suspensions in assay tubes are not simply related to numbers of cells. However, this does not mean that the optical properties are any less useful as a measure of response in assay work. Turbidity of cell suspensions may be measured either by light scattered using a suitable nephelometer or by monochromatic light transmitted in an absorptiometer or spectrophotometer.

The EEL nephelometer has been much used in Britain in turbidimetric assay. In this instrument, light from a tungsten filament lamp is passed first through a suitable color filter upward through the bottom of a standardized test tube containing the suspension. Scattered light is reflected by a parabolic mirror to an annular-shaped barrier-type photocell surrounding the base of the tube. Suspensions are compared with an arbitrary standard consisting of a Perspex rod having an etched surface. This rod has a diameter just slightly less than the internal diameter of the standard test tubes into which it fits neatly. Water or other inert liquid may be placed in the tube to fill the gap between the Perspex and the glass. The apparent turbidity of the standard may be modified by using liquids of different refractive index if needed. However, it is generally convenient to work with water. The instrument is set to give a reading of 100 with the Perspex standard, and turbidities of suspensions are compared and recorded in arbitrary units.

It was shown by Hewitt (1975) that instrument response was directly proportional to cell suspension in the case of a formalin-treated overnight culture of *Staphylococcus aureus* cells in the range of concentrations that would be encountered in routine assays. When an absorptiometer is used, the measured response may be either percentage of the monochromatic light transmitted (%T) or absorbance (A). The latter is defined by the expression:

$$A = \log(I/T) \tag{5.4}$$

where I is the intensity of light incident on the absorbing layer and T is the intensity of light emerging from the absorbing layer.

The instrument, of course, actually measures the light that is transmitted whether the response is shown as %T or as A. Absorbance is the more logical parameter to work with as it can be approximately proportional to the true assay response. The absorbance found depends on the following factors: the absorbancy characteristics of the cells, the length of the light path through the suspension, and the wavelength of the monochromatic light.

The same factors also apply to the measurement of absorbance by substances in solution. However, in the case of cell suspensions the reduction in intensity of light is not only by absorbance but also by light scattering. Some of the light is scattered in a generally forward direction and so may reach the detector. If the proportion of forward-scattered light actually reaching the detector can be reduced, then the observed absorbance will be greater. The detector's acceptance angle for light is dependent on the geometry of the optical system. It may be reduced by interposing a mask between the cuvette containing the cell suspension and the detector window as described by Kavanagh (1972).

The relationship between absorbance of monochromatic light and concentration of absorbing substance is expressed by the Beer–Lambert law as

$$A = kc \qquad\qquad (5.5)$$

in which c is concentration and k is a constant. In other words, absorbance is directly proportional to concentration of absorbing material for a fixed thickness of absorbing layer.

While it is not the purpose of this book to discuss absorptiometry in detail, a few pointers are appropriate. The Beer–Lambert law represents an ideal situation which may be attained or approached in practice according to circumstances. In practice the light used in any absorptiometer is not truly monochromatic but has a finite bandwidth. The width of band is generally expressed as half-bandwidth, which is its width at half the peak height as illustrated in Fig. 5.10 for a typical filter and for a monochromator such as that incorporated in a high-precision spectrophotometer. In general, conformity with the Beer–Lambert law is more likely to be approached when using a narrow half-bandwidth. It can be shown from theoretical considerations that the best precision of optical absorbance measurements is attained when %T is about 40, corresponding to an absorbance of about 0.4. A useful working range on either side of this ideal is

$$\%T = 20 \text{ to } 60$$

corresponding to

$$A = 0.70 \text{ to } 0.22$$

although with the more elaborate spectrophotometers precise measurements are achieved in a wider range.

If deviations from the Beer–Lambert law are substantial it is a simple, practical matter to prepare a calibration curve or chart to relate observed optical absorbance to relative cell concentration (RCC) or adjusted optical density (AOD); both of these terms are used by different workers to denote the same correction procedure. From a formalin-treated cell suspension a series of dilutions is prepared accurately to correspond with the cell concentrations that would be encountered in a routine assay. Absorbance of the various dilutions is measured at the wavelength to be used in assays. A graph is then prepared by plotting observed absorbance against RCC. The graph may then be used to convert observed absorbance to RCC. However,

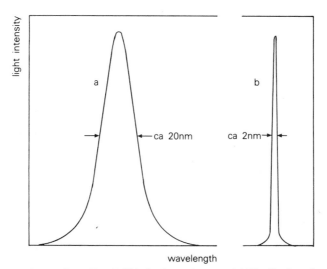

Fig. 5.10.   An illustration of bandwidths in absorptiometry. (a) Distribution of wavelengths typical of a filter for a monochromator of a simple spectrophotometer; the width of the band of light might be 10–30 nm at half the peak height. (b) Distribution of wavelengths of light typical of the monochromator of a high-precision spectrophotometer; the bandwidth may be 1 nm or less at half the peak height.

if the correction procedure is to be used frequently it would be more convenient to interpolate and prepare tables relating the two parameters or to incorporate the conversion in a computer program.

It may be necessary to prepare separate calibration curves for different test organisms, particularly if they differ greatly in shape, size, or color. Experimentation will show whether this is necessary in individual situations.

As cell suspensions reduce the transmittance of light partly by scattering, a wide range of wavelengths could be used. It is suggested by Kavanagh (1972) that wavelengths between 550 and 650 nm are suitable, because absorbance is adequate in this range and absorbance by the medium itself is low. The medium may absorb strongly at 400 nm. The change in absorbance with wavelength in the case of a suspension of *S. aureus* is shown in Fig. 5.11, which is from Rippere (1979). This shows a portion of a series of response curves in which absorbance is plotted against dose on a log scale. It is seen that absorbance, and also to a slight extent slope of response line, increase as wavelength of light is reduced from 650 to 530 nm. If the test organism is colored, then its absorption peak may be the most appropriate

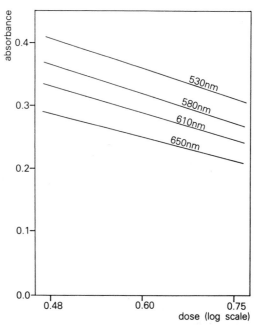

Fig. 5.11. Influence of wavelength in the measurement of absorbance by suspensions of microorganisms *(Staphylococcus aureus)*. Absorbance, and also to a slight extent slope of response line, increase as wavelength of light decreases.

wavelength at which to measure. If the sample contains any colored material in sufficient concentration to absorb appreciably in the test preparation, then its absorption band should be avoided.

In a properly conducted assay, conditions and techniques are such that the *actual* response reflects faithfully the dose of specific biological activity in each tube. It is necessary to measure the individual responses with due care and awareness of possible sources of error so that *observed* responses also reflect faithfully the doses of specific biological activity. It is also desirable, particularly in a busy laboratory, that measurements be carried out expeditiously. Turbidities may be read by placing the assay tube directly in the nephelometer where it serves as a cuvette. This is also possible with some simple absorptiometers, even though the geometry of the tube is not really appropriate for absorption measurements. Naturally, if assay tubes are to be used as cuvettes in any optical instrument, they must be of uniform shape and size, and of the same quality, color, and thickness of glass.

To make a measurement of optical properties using the actual assay tube, the tube's external surface must be dried then polished with a soft cloth or tissue. The suspension must be agitated to ensure that cells are properly dispersed.

Possible sources of error are insufficient agitation leading to inadequate dispersion of cells, too much agitation resulting in the production of tiny bubbles which increase the scattering of light, and too rapid reading so that the instrument fails to register the correct response.

It is necessary to establish that readings are reproducible by repeating the agitation and measurement of optical properties after a short interval. Naturally, the instrument itself must demonstrably give reproducible responses.

A problem specific to rod-shaped organisms is reported by Kavanagh (1963). Suspensions of rod-shaped organisms exhibit flow birefringence, so that when there is turbulent flow or eddy currents, as in a freshly shaken tube or a cuvette just after filling with the suspension, absorbance measurements fluctuate until the organism has returned to its completely random orientation. This problem is obviated when using a flow cuvette under standard conditions which ensure that the orientation of the rods is sufficiently uniform to eliminate detectable fluctuations in the absorbance measurement. Such a system is employed in the automatic reading module of the Autoturb, which is described in Section 5.6 of this chapter. This problem of flow birefringence has not been observed by the authors when using the EEL nephelometer, presumably because of the different geometry of the instrument in which light scattered through a wide range of forward angles is reflected by a parabolic mirror to the detector.

The importance of objectivity in observing responses was stressed in Chapter 4 with reference to the agar diffusion assay. This principle is equally important in the measurement of turbidity or any other response. Anticipation of the desired observation must be avoided at all costs. Automation of reading is a certain way to eliminate such anticipation.

5.5.4   Calculation of the Potency Estimate

The calculation methods that may be used in any individual assay are dependent on the experimental design and the nature of the dose–response relationship. Experimental design is discussed in Chapter 8, and so it will suffice to state here that a symmetric design is inherently more efficient than any asymmetric design. The use of the latter may be justified, however, when the analyst has no prior knowledge of the probable region of the potency of the sample and so may set up the assay using a standard curve covering a good range of potencies and using two or more nominal potencies for the "unknown," hoping that at least one of them will lead to responses in the range covered by the standard curve. Even though symmetric assays are the logical choice in the majority of circumstances, there is widespread use of standard-curve assays. When graphs are drawn, these should generally be the best smooth curve or the best straight line to pass close to all points. Interpolation from strongly curved regions at the extremities of the graph is best avoided. Any sudden change in direction in the central region of the graph should be suspect and may be due to some gross error such as a dilution mistake or use of the wrong test solution.

Some workers draw point-to-point graphs joining the mean responses to individual doses. However, the individual points and their means are subject to random error, and so it is better to smooth out these errors by drawing a curve of best fit.

When the mean responses plotted against either dose or logarithm of dose appear to represent a straight line, then the position and slope of the best straight line corresponding to the points may be calculated. A commonly used calculation is that of the U.S. CFR, which requires that doses be equidistant on the scale on which they are plotted in the graph. The calculation is carried out as follows:

Let the mean responses to the five doses be S_1, S_2, S_3, S_4, and S_5, where S_1 is the response to the lowest, and S_5 is the response to the highest dose. From these raw data two values H and L are calculated; these represent the best estimate of response to high and low doses, respectively, that can be calculated from all the data. They are calculated thus:

$$H = \tfrac{1}{5}[3S_5 + 2S_4 + S_3 - S_1] \qquad (5.6)$$

$$L = \tfrac{1}{5}[3S_1 + 2S_2 + S_3 - S_5] \qquad (5.7)$$

A dose–response line is then obtained by plotting the values H and L against dose (or more commonly logarithm of dose) and joining the two points. Sample potencies may then be estimated by interpolation from the standard line.

The use of Eqs. (5.6) and (5.7) is illustrated by their application in Example 5.1, an assay of carbenicillin described by Stankewich and Upton (1973).

Example 5.1: Assay of Carbenicillin

Test organism: E. coli (51A266)
Incubation: 3.25 hours at 37°C
Standard dose range: 0.5–0.9 μg/ml in final broth
Observations: Absorbance at 530 nm using an Autoturb reader module equipped with a model 330 Turner spectrophotometer and a 1-cm flow cell

Table 5.3

Assay of Carbenicillin: Mean Responses to a
Range of Concentrations of Reference Standard
(Example 5.1)

Dose of carbenicillin (μg/ml)	Mean absorbance at 530 nm
0.50	0.425
0.55	0.385
0.60	0.358
0.65	0.320
0.70	0.282
0.75	0.233
0.80	0.200
0.85	0.159
0.90	0.139

Doses and corresponding means of replicate responses are shown in Table 5.3. A plot of the five mean responses against doses ranging from 0.6 to 0.8 μg/ml approximated a straight line and so is used here for the purpose of demonstrating the calculation based on five points. (Note that in actual routine assays, Stankewich and Upton used only dose levels 0.6, 0.7, and 0.8 μg/ml and did not use this calculation procedure.) The values of H and L were calculated thus:

$$H = \tfrac{1}{5}[(3 \times 0.200) + (2 \times 0.233) + (0.282) - (0.358)] = 0.1980$$

$$L = \tfrac{1}{5}[(3 \times 0.358) + (2 \times 0.320) + (0.282) - (0.200)] = 0.3592$$

The graph of Fig. 5.12 was drawn by joining the two points plotted for H and L. The means of the observed responses are also plotted for comparison.

Potencies of sample test solutions may be interpolated from the straight-line graph. Thus, the potency of a test solution giving a mean response of absorbance 0.355 would be seen from the graph to be 0.63 μg/ml.

When the response, or some mathematical transformation of the response, plotted against logarithm of dose approximates to a straight line, sample potency can be estimated by a purely arithmetic procedure as in the case of the plate assays that were described in Chapter 4. It is necessary for the application of these formulas that doses form a geometric progression so that their logarithms are uniformly spaced. The expressions for the calculation of E and F are based on the same principles that were illustrated in Section 4.6.6 for a symmetric three-dose level assay. These expressions for symmetric assays of several dose levels are presented in Appendix 6.

The calculation procedure for a six-dose level symmetric assay in which means of triplicate responses have been transformed to probits is illustrated in Example 5.2. The effectiveness of the probit in straightening the dose–response line may be seen by comparing Fig. 5.13a and Fig. 5.13b, in which the data of Example 5.2 are plotted first as mean absorbance (Fig. 5.13a),

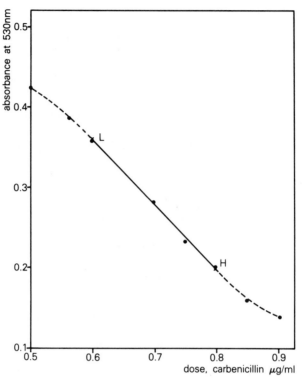

Fig. 5.12. A standard dose–response line for carbenicillin in the tube assay using *Escherichia coli* as test organism. The plot of mean response (absorbance at 530 nm) against dose approaches a straight line most closely in the range 0.6–0.8 μg/ml. The graph illustrates Example 5.1. (●) Mean of observed responses.

then as probit of proportionate response against logarithm of dose (Fig. 5.13b). Although the plot of probits also shows some slight degree of sigmoid character, it has been shown that this is quite negligible in its influence on the potency estimate in this symmetric assay.

For other examples of the calculation of potency estimates from this type of assay, the reader may consult the work of Hewitt (1977).

Example 5.2: Assay of Tetracycline

Test organism: *S. aureus* ATCC 6538-P
Incubation: 3.5 hours at 37°C
Observations: Transmittance at 530 nm using a Spectronic 20 spectrophotometer and a 1-cm cuvette

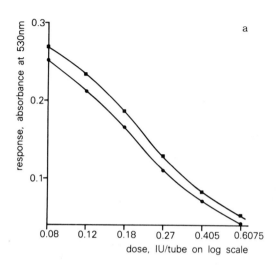

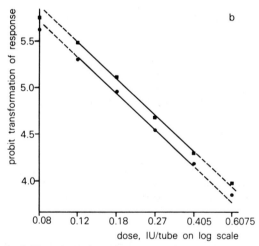

Fig. 5.13. Standard (●) and sample (■) dose–response lines in an assay of tetracycline using *Staphylococcus aureus* as test organism. (a) Mean response plotted as absorbance at 530 nm against logarithm of dose; the resultant lines are distinctly sigmoid. (b) Probit of proportionate response plotted against logarithm of dose; the resultant lines are straight along almost their entire length.

The observations and their transformation to probits are shown in Table 5.4.

The calculation then proceeds using the probits of proportionate response, which are designated S_1 to S_6 and T_1 to T_6 for standards and unknowns, respectively.

From Appendix 6 it is seen that E and F for this assay design are obtained as

$$E = \tfrac{1}{70}[5(S_6 + T_6) + 3(S_5 + T_5) + (S_4 + T_4) - (S_3 + T_3) - 3(S_2 + T_2) - 5(S_1 + T_1)]$$

and

$$F = \tfrac{1}{6}[(T_6 + T_5 + T_4 + T_3 + T_2 + T_1) - (S_6 + S_5 + S_4 + S_3 + S_2 + S_1)]$$

Substituting the probits from Table 5.4, E and F become

$$E = \tfrac{1}{70}[39.135 + 25.470 + 9.241 - 10.081 - 32.355 - 57.100]$$
$$= -0.367$$

$$F = \tfrac{1}{6}[29.353 - 28.491] = 0.1438$$

$$b = \frac{E}{I} = -0.3670/\log 1.5 = -0.3670/0.17609 = -2.08415$$

$$M = F/b = 0.1438/-2.08415 = -0.0690$$

Potency ratio is given by the antilogarithm of M, which is 0.8531.

For an explanation of the symbols used in this example see Chapter 4, Section 4.6.6.

5.6   Automated Methods

It will be clear from Section 5.5 that the tube assay for growth-inhibiting substances involves much routine and repetitive work. If many samples are to be assayed, then the manual work load becomes very heavy and the probability of failure to maintain the necessary standards of technique increases. In consideration of the critical factors described in Section 5.4, it is evident that the quality of assays may drop to an unacceptable level. Moreover, a large quantity of volumetric and other glassware is used, placing a heavy burden on cleaning services.

There are obvious potential advantages in automation of some or all of the steps in an assay, so that the tedium of the work may be reduced and conditions of assay so standardized as to lead to acceptable reproducibility of potency estimates.

Several automated systems have been described in the literature, having been designed by analysts and engineers for use in their own laboratories. The steps which are amenable to automation are dilution of the sample and filling of the tubes with dosed inoculated nutrient medium, and sampling and reading the optical absorbance of the test preparations at the end of incubation. Systems incorporating these features have been described by Kuzel and Kavanagh (1971a,b) and by Berg *et al.* (1975). An automatic dilution system is described by Palmer and Hamilton (1975). The Autoturb

Table 5.4

Assay of Tetracycline (Example 5.2)[a]

Dose (IU/tube)	Log of 100 × dose	Transmittance	Mean transmittance	Mean absorbance A	A/A_{max}	Probit	Designation
Standard							
0.0800	0.903	56.5, 56.0, 55.5	56.0	0.252	0.737	5.634	S_1
0.1200	1.079	62.0, 60.5, 62.0	61.5	0.212	0.620	5.306	S_2
0.1800	1.255	68.5, 68.0, 68.5	68.3	0.166	0.485	4.962	S_3
0.2700	1.431	77.5, 77.5, 77.5	77.5	0.111	0.325	4.546	S_4
0.4050	1.608	85.0, 85.0, 85.0	85.0	0.071	0.208	4.187	S_5
0.6075	1.784	90.5, 90.5, 91.0	90.7	0.043	0.126	3.855	S_6
Sample							
0.0800	0.903	54.0, 54.0, 54.0	54.0	0.268	0.784	5.686	T_1
0.1200	1.079	58.0, 58.5, 58.5	58.3	0.234	0.684	5.479	T_2
0.1800	1.255	65.0, 65.0, 65.0	65.0	0.187	0.547	5.118	T_3
0.2700	1.431	74.5, 74.0, 74.0	74.2	0.130	0.380	4.695	T_4
0.4050	1.608	82.5, 82.5, 83.0	82.7	0.083	0.243	4.303	T_5
0.6075	1.784	89.0, 89.0, 88.5	88.8	0.052	0.152	3.972	T_6
Zero control	—	45.5, 45.0, 46.0	45.5	0.342	1.000		

[a] Observations, transmittance at 530 nm using a Spectronic 20 spectrophotometer, together with transformation of these responses to the corresponding probit.

system developed by Kuzel and Kavanagh is commercially available and is manufactured by Elanco Products Company of Indianapolis, Indiana.

A notable feature of the Autoturb system is the high-precision incubator bath, for which it is claimed that temperature does not differ by more than $0.02\,°C$ between any two points in the bath. The value of this precise control is clear in consideration of the effect of temperature change as discussed in Section 5.4. Both the Autoturb system and the system described by Berg employ a flow cuvette, and both recognize the value of this in eliminating the problems of birefringence that were mentioned in Section 5.5.3. Although such automated equipment may seem expensive as compared with the small investment so frequently made in microbiological assaying (Hewitt, 1982), it is not expensive when compared with the physicochemical instrumentation that is so widely accepted. It is concluded that investment in automated or semiautomated equipment for microbiological assay would be well justified in a busy laboratory.

References

Berg, Th. M., Den Burger, J. M., and Behagel, H. A. (1975). *In* "Some Methods for Microbiological Assay" (R. G. Board and D. W. Lovelock, eds.). Academic Press, New York.
Bliss, C. I. (1934a). *Science* **79**, 38.
Bliss, C. I. (1934b). *Science* **79**, 409.
Brown, M. R., and Garrett, E. R. (1964). *J. Pharm. Sci.* **53**, 179.
Cooper, K. E. (1963). *In* "Analytical Microbiology" (F. W. Kavanagh, ed.), Vol. I. Academic Press, New York.
Fisher, R. A., and Yates, F. (1963). "Statistical Tables for Use in Biological, Agricultural and Medical Research," 6th ed. Longman, London.
Gaddum, J. R. (1933). *Med. Res. Council Spec. Rep. Ser.,* No. 103.
Garrett, E. R., and Miller, G. H. (1965). *J. Pharm. Sci.* **54**, 427.
Hemmingson, A. M. (1933). *Q. J. Pharm. Pharmacol.* **6**, 39, 187.
Hewitt, W. (1975). Unpublished observations.
Hewitt, W. (1977). "Microbiological Assay." Academic Press, New York.
Hewitt, W. (1982). *Pharm. Int.* **3**, 11, 370.
Hewitt, W. (1987). Unpublished observations.
Kavanagh, F. W., ed. (1963). "Analytical Microbiology," Vol. I. Academic Press, New York and London.
Kavanagh, F. W. (1968). *Appl. Microbiol.* **16**, 777.
Kavanagh, F. W., ed. (1972). "Analytical Microbiology," Vol. II.
Kavanagh, F. W. (1975). *J. Pharm. Sci.* **64**, 844.
Kuzel, N. R., and Kavanagh, F. W. (1971a). *J. Pharm. Sci.* **60**, 764.
Kuzel, N. R., and Kavanagh, F. W. (1971b). *J. Pharm. Sci.* **60**, 767.
Palmer, G. H., and Hamilton, P. (1975). *In* "Some Methods for Microbiological Assay" (R. G. Board and D. W. Lovelock, eds.). Academic Press, London.
Rippere, R. A. (1979). *J. Assoc. Off. Anal. Chem.* **62**, 4, 951.
Stankewich, J. P., and Upton, R. P. (1973). *Antimicrob. Agents Chemother.* **3**, 3, 364.

CHAPTER 6

TUBE ASSAYS FOR GROWTH-PROMOTING SUBSTANCES

6.1 Introduction

The general pattern of growth of a microorganism in a liquid nutrient medium was outlined in the introduction to Chapter 5. It is equally relevant to the assay of growth-promoting substances by the tube assay. Again, for convenience and brevity, the term tube assay will be used to refer only to the assay in a liquid medium and not the tube version of the agar diffusion assay. The response measured is generally either the turbidity arising from the growth of cells or the acid produced by the cells, which is titrated with alkali. The principle of the tube assay for a growth-promoting substance is thus as described below.

A suitable organism is provided with a liquid nutrient medium that contains an adequate quantity of all but one of the substances that are essential for its growth. Addition of that one unique substance will permit growth of the organism; the extent of the resulting growth will be related quantitatively to the amount of the unique substance added. In practice the situation is not generally so ideal. Instead of only one substance being capable of completing the medium and thus permitting growth of the test organism, there may be a family of related substances having different potencies as growth promoters. Thus, growth-promoting substance assays are not necessarily specific for a single chemical entity. Also, the requirement may not be absolute; in some cases growth may continue, but at a reduced rate after depletion of the "essential" substance from the medium. Despite these deviations from the ideal, growth-promoting substance assays are extremely useful methods provided that their limitations are understood and taken into consideration.

6.2 Outline of the Method

The tube method for growth-promoting substances is generally carried out using a total of 10 ml of liquid medium in each tube. To set up the assay, graded dilutions of both the standard reference substance and the sample to be tested are added to a series of tubes; e.g., 1 ml, 2 ml, and 3 ml or more of each test solution may be used. Accurately measured volumes of water are then added to each tube to make a total of 5 ml. Additionally, for each series, two tubes are set up containing 5 ml of water; of these, one will serve as an inoculated blank and one as an uninoculated blank. A few extra tubes of high-dose standard are also included solely for checking for the end of growth; these are not to be used in the computation of potency estimate. Each series of tubes is prepared with a replication of perhaps three or four tubes for each dose level of each preparation. To each tube is then added 5 ml of the specific nutrient medium for this assay, which has been prepared at double strength. The specific medium is deficient in the substance to be assayed but complete in all other nutrients. The tubes are then subjected to heat treatment to kill any contaminating organisms; after cooling, each tube is inoculated with one drop of a suspension of the specific test organism. The tubes are incubated, usually for 16 hours or more if the response is to be measured turbidimetrically but for longer periods such as 72 hours if the response is to be measured titrimetrically. Potencies of the unknown preparations are then estimated by comparison of sample and standard responses either graphically or arithmetically.

6.3   Nature of the Response

Mathematical treatments of assay responses sometimes assume the ideal that response is directly proportional to dose of growth-promoting substance, then recognize that the ideal is not always attained and so incorporate statistical methods to measure the significance of deviations, e.g., the significance of curvature of the dose–response lines. It is pertinent to inquire into the true patterns of both growth and acid production in the conditions of an assay tube and how deviations from the assumed straight-line dose–response relationship arise. This will be done first for turbidimetric assays by looking at growth patterns, then for titrimetric assays considering both the growth pattern and the rate of acid production. Initially some theoretical models will be proposed, then some practical observations will be considered. The growth-promoting substances that are assayed microbiologically generally fall into one of two groups: amino acids, and vitamins of the B group. The former are the building blocks from which the cell proteins are formed; the latter combine with other substances to form the coenzymes which are essential for cell metabolism and growth.

6.3.1   Amino Acids As Limiting Factors: Theory

If a particular amino acid were a limiting factor in the sense of providing the unique and essential building blocks for the growth of each cell, then clearly in an assay tube the total number of cells produced by the time that growth ceases must be directly proportional to the number of molecules of that essential amino acid. This reasoning assumes, of course, that all other essential growth substances are present in adequate excess and that no other factors modify the course of growth. For assay purposes it is necessary to consider the time taken to attain maximum growth for different levels of the limiting factor. Monod (1949) observed that the relationship between growth rate and nutrient concentration could be described by an equation of the form:

$$\mu = \frac{\mu_{\max} S}{K_S + S} \tag{6.1}$$

where μ is the specific growth rate, $\mu_{\max}$, maximum specific growth rate, S, substrate concentration, and K_S, a constant which is equal to the substrate concentration when $\mu = 0.5\mu_{\max}$.

It seems clear that Eq. (6.1) could be applied to each individual nutrient substance, thus leading to a complex expression describing growth rate. However, as stated in Section 6.1, it is a requirement of a growth-promoting substance tube assay that each individual nutrient substance in the medium is present in adequate quantity. We may define "adequate" as "in such concentration that when growth has ceased, its concentration remains many times greater than its own value of K_S." In these circumstances the Monod equation when applied only to the substance being assayed (the limiting substance) may be expected to describe the growth rate in an assay tube.

It will be shown that during the course of cell growth over several generations there remains a substantial excess of the limiting factor for the greater part of the period of growth. It is only toward the end, when cell population would tend to be increasing most rapidly (on an arithmetic scale), that the residual concentration of limiting substance drops suddenly, thus reducing the specific growth rate sharply and signaling the end of growth that has been almost exponential.

In the application of the Monod model shown here, the following arbitrary assumptions are made:

(i) For the growth of each cell, p molecules of the specific limiting amino acid A are required.

(ii) The inoculum level in all tubes is n cells/tube.

(iii) Growth ends after about 10–11 generations according to the limiting number of molecules of A in the tube.

(iv) The limiting quantity of A in tubes at dose level 1 is $2^{10} \times np$

molecules $= 1024np$ molecules, which is sufficient to permit an increase of $1024n$ cells per tube. Thus the final number of cells in a tube should be $1025n$.

(v)   Dose levels 2 and 3 correspond to $2048np$ and $3072np$ molecules/tube, respectively, leading to final cell concentrations of $2049n$ and $3073n$ per tube.

(vi)   The constant K_S (which is naturally the same for all dose levels) is equal to $0.1 \times$ the concentration of limiting amino acid at dose level 1 (i.e., $102.4np$).

(vii)   The generation time when A is abundant (i.e., when $\mu = \mu_{max}$) is t minutes.

It follows that at dose level 1 the initial specific growth rate is given by

$$\mu = \mu_{max} \left[\frac{1024np}{(102.4 + 1024)np} \right] = 0.90909$$

If it were assumed that the specific growth rate remained constant during the first t minutes of growth, then the number of cells would increase from n to $1.9091n$ during that period. It follows that the number of molecules of A remaining in the substrate would have dropped from $1024np$ to $(1024 - 0.9091)np = 1023.0909np$ per tube. Thus, for the growth period t to $2t$ a new value of specific growth rate would be calculated as

$$\mu = \mu_{max} \left[\frac{1023.0909np}{(102.4 + 1023.0909)np} \right] = 0.90902\mu_{max}$$

Iterative calculations such as these show that changes in the value of μ are barely perceptible during the first few generations and that μ is still greater than 0.9 even after $7t$ minutes of growth. After this, μ drops at an increasingly rapid rate and it becomes no longer realistic to assume the approximation that it remains constant during t minutes of growth. It is necessary to introduce a modified procedure to recalculate μ after fractional time periods such as $0.1t$ and even less as the end of growth approaches. With the aid of a computer it is a simple matter to use the modified procedure for calculations, and so a time period of $0.01t$ was used to calculate the values for specific growth rate (μ) and relative cumulative cell concentration (RCCC) that are presented in Table 6.1. (It will be noted that there are some large blank areas in this table; values of μ and RCCC have been omitted in these areas so as to highlight the areas of greater interest.)

In addition to the three dose levels, the calculation was also applied to a much lower concentration of limiting substance (10% of that of dose level 1) to represent the case when there is a trace of the limiting substance present,

Table 6.1

Growth of an Organism in Substrates Having Different Levels of a Limiting Amino Acid[a]

Time Units	Dose level "trace"		Dose level 1		Dose level 2		Dose level 3		Exponential growth, RCCC
	μ	RCCC	μ	RCCC	μ	RCCC	μ	RCCC	
0	0.5000	1.0000	0.9091	1.0000	0.9524	1.0000	0.9677	1.0000	1
1.0	0.4988	1.4944	0.9091	1.9091	0.9524	1.9524	0.9677	1.9677	2
2.0	0.4970	2.2461	0.9089	3.6433	0.9524	3.8118	0.9677	3.8720	4
10.0	0.3589	46.228	0.8098	606.80	0.9274	791.86	0.9567	863.93	1024
11.0	0.2920	61.328	0.3049	993.20	0.8524	1503.2	0.9324	1681.5	2048
11.2	**	**	0.0701	1020.7	0.7975	1694.6	0.9202	1917.1	2353
11.4	**	**	0.0058	1024.7	0.6686	1890.2	0.8997	2181.9	2702
11.6	**	**	0.0004	1025.0	0.2687	2035.5	0.8597	2475.6	3104
11.8	**	**	*	*	0.0001	2049.0	0.7553	2788.1	3566
12.0	0.2078	76.676	*	*	*	*	0.2961	3045.7	4096
12.2	**	**	*	*	*	*	0.0002	3073.0	4705
14.0	0.0589	97.042	*	*	*	*	*	*	
16.0	0.0094	102.44	*	*	*	*	*	*	
18.0	0.0013	103.27	*	*	*	*	*	*	
20.0	0.0002	103.38	*	*	*	*	*	*	
21.0	0.0001	103.39	*	*	*	*	*	*	
22.0	0.0001	103.40	*	*	*	*	*	*	

[a] Values of specific growth rate (μ) and relative cumulative cell concentration (RCCC) are as calculated from the Monod equation [Eq. (6.1)]. The parameters used are $K_s = 102.4$; $S = 1024$, 2048, and 3072 for dose levels 1, 2, and 3 of limiting amino acid, respectively; $S = 102.4$ to represent an adventitious trace of limiting amino acid. Exponential growth is shown for comparison. *, Values of μ and of RCCC have not been calculated in these cases, as μ is close to zero and RCCC is close to its maximum value. **, Values of μ and RCCC in these cases would be superfluous.

as may happen when there is some carry-over in the inoculum or when the medium is not completely free of the limiting substance.

Values of μ and RCCC at selected time periods are shown in Table 6.1. Increase in cell concentration with time is plotted on a log scale in Fig. 6.1. From both the table and the graph it may be seen that at higher dose levels μ is high and growth is close to exponential at first, then μ drops suddenly. At increasing dose levels an increasingly abrupt end is forecast by these applications of the Monod equation. In contrast, at the very low levels of essential amino acid corresponding to an unwanted trace, it appears that the decrease in growth rate with time should be very gradual. The data of Table 6.1 are also used to plot the graphs of Fig. 6.2, i.e., cell concentration versus

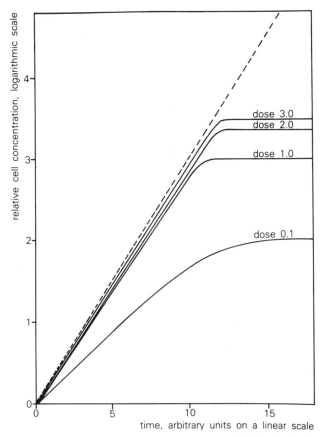

Fig. 6.1. A representation of cell growth when limited by differing levels of one essential amino acid. These theoretical curves are drawn from the values in Table 6.1 that were calculated by means of the Monod equation.

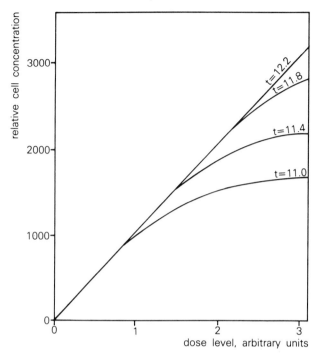

Fig. 6.2. A representation of the forms of dose–response lines that would be expected at different growth periods when cell growth is limited by one essential amino acid. These theoretical curves are drawn from the values in Table 6.1 that were calculated by the Monod equation.

dose level for various growth periods. These lines correspond to the dose–response lines of assays, and they predict that response will approach direct proportionality to dose only if there is adequate incubation time. They also suggest that slightly longer incubation times will be necessary at higher dose levels.

6.3.2 Amino Acids as Limiting Factors: Practice

Some examples of practical evidence as to the nature of growth when there is a limiting concentration of a particular amino acid are provided by the work of Toennies and Shockman (1953), who studied the growth of *Streptococcus faecalis* in conditions of single-amino acid limitation. They showed that in the cases of limiting quantities of L-histidine, L-isoleucine, L-leucine, L-threonine, and L-valine, growth appeared to proceed exponentially up to the "depletion point," i.e., the time at which all the limiting amino acid had been taken up by the cells and none remained in the liquid phase of the suspension. Growth still continued beyond this point but at a reduced rate. The growth was followed by measurement of optical absorbance.

In one experiment, in which L-valine was the limiting amino acid, changes in the dry weight of cells were also measured. It was shown that both optical absorbance and dry weight of cells increased by about 40% in the 12-hour period of postexponential growth. In the case of L-histidine, postexponential growth was followed by lysis as shown by a reduction in turbidity. In the case of L-lysine, however, rapid lysis followed almost immediately after the end of exponential growth.

The growth curves for different limiting values of L-valine are shown in Fig. 6.3. These differ from the theoretical curves of Fig. 6.1 in that growth continues steadily far beyond the end of exponential growth. However, significant points of similarity are (1) the near coincidence of growth rate in the phases that appear to be exponential, regardless of the concentration of

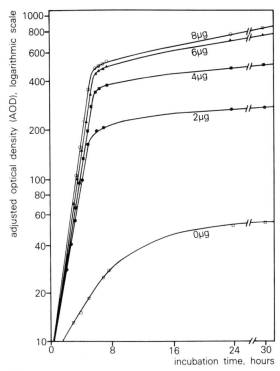

Fig. 6.3. Practically determined growth curves for *Streptococcus faecalis* (9790) when growth is limited by differing levels of L-valine. From the work of Toennies and Shockman (1953).

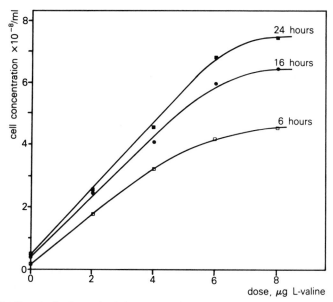

Fig. 6.4. Practically determined dose–response lines after different growth periods in the assay of L-valine using *Streptococcus faecalis* (9790). These are drawn from the same data of Toennies and Shockman (1953) as were used for Fig. 6.3.

amino acid at dose levels 2–8 μg, and (2) the very much more gradual reduction in growth rate in the cases of the curves representing an adventitious trace of amino acid.

Cell concentrations are plotted against concentration of L-valine corresponding to different periods of incubation in Fig. 6.4. These confirm what was deduced from Fig. 6.2, that curvature of the dose–response line decreases with increasing length of incubation. At 16 hours the dose–response line was linear up to 6 μg.

6.3.3 Vitamins as Limiting Factors: Theory

It was stated in Section 6.1 that vitamins of the B group combine with other substances to form the coenzymes that are essential for cell metabolism and growth. This suggests that vitamins could play a growth rate-determining role. However, it is reasoned that in the conditions of an assay tube there would be an excess of the vitamin during the early and middle hours of growth and that the rate-regulatory function would become operable only in the latter part of the period of incubation. The organism would multiply in a true log phase throughout the greater part of the period of growth until the number of cells had become large relative to the number of molecules of vitamin available to facilitate its continued growth. At this stage the vitamin would become the factor limiting the rate of growth, and so a change from logarithmic to arithmetic growth would be expected. The rate of arithmetic growth would be expected to be directly proportional to concentration of vitamin in individual tubes. This is illustrated by a mathematical model which makes some simplifying assumptions as follows:

(i) The inoculum consists of n cells in the logarithmic phase of growth.

(ii) Relative dose levels of the limiting vitamin are in the ratios $1:2:3:4$.

(iii) The vitamin has a catalytic role and is not consumed.

(iv) The concentration of vitamin at dose level 1 becomes a limiting factor after exactly 12 generations, when the number of cells has reached $2^{12} \times n = 4096n$.

(v) The change from logarithmic to arithmetic growth is abrupt.

It is a logical extension of assumption (iv) that concentrations of vitamin at dose levels 2 and 4 would become limiting factors after 13 and 14 generations, respectively. It can also be calculated that the vitamin concentration at dose level 3 would become a limiting factor after 11.585 generations. These data, together with cell concentration at point of change from logarithmic to arithmetic growth, are presented in Table 6.2.

The numbers of cells per tube calculated on the basis of these assumptions for different dose levels and durations of growth spanning both

Table 6.2

Growth of an Organism as Limited by Different
Levels of Vitamin[a]

Dose level	Number of generations	Number of cells
1	12	4096n
2	13	8192n
3	13.585	12288n
4	14	16384n

[a] It is first supposed that at dose level 1 exponential growth ceases after 12 generations and thereafter growth rate is limited by vitamin concentration. Numbers of cells and numbers of generations corresponding to cessation of exponential growth at four dose levels are shown. An inoculum of n cells is assumed.

logarithmic and arithmetic phases are shown in Table 6.3, for which purpose it is further assumed that $n = 10,000$. From the calculated numbers of cells per tube that are summarized in Table 6.3, "growth curves" have been drawn. These are shown in Figs. 6.5 and 6.6 with the numbers of cells per tube plotted on arithmetic and logarithmic scales, respectively. The numbers of cells per tube corresponding to the varying doses of vitamin and numbers of growth periods are summarized in Table 6.4 together with the "response ratios," i.e., the ratios of "responses" to different dose levels for each number of growth periods. These same data are also expressed graphically in Fig. 6.7, where cells per tube is plotted against dose of vitamin for the varying numbers of growth periods.

It is thus deduced that (i) response ratios will approach more closely the ideal straight-line relationship as incubation time increases, and (ii) response ratios will approach more closely the ideal at the lower end of the dose range.

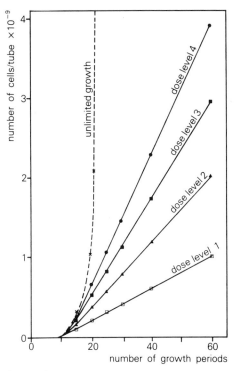

Fig. 6.5. A series of theoretical curves representing the growth of a test organism according to differing limiting doses of an essential vitamin. These are based on the calculated values of Table 6.3. Growth is shown on an arithmetic scale.

Table 6.3

Further Representation of Growth of an Organism with Limiting Levels of a Vitamin[a]

Growth period	Dose level 1		Dose level 2		Dose level 3		Dose level 4	
	CGTI × 10⁻⁷	CCC × 10⁻⁷	CGTI × 10⁻⁷	CCC × 10⁻⁷	CGTI × 10⁻⁷	CCC × 10⁻⁷	CGTI × 10⁻⁷	CCC × 10⁻⁷
Inoculum		0.001		0.001		0.001		0.001
0–1	0.001	0.002	0.001	0.002	0.001	0.002	0.001	0.002
1–2	0.002	0.004	0.002	0.004	0.002	0.004	0.002	0.004
2–3	0.004	0.008	0.004	0.008	0.004	0.008	0.004	0.008
3–10	1.016	1.024	1.016	1.024	1.016	1.024	1.016	1.024
10–11	1.024	2.048	1.024	2.048	1.024	2.048	1.024	2.048
11–12	2.048*	4.096	2.048	4.096	2.048	4.096	2.048	4.096
12–13	2.048	6.144	4.096*	8.192	4.096	8.192	4.096	8.192
13–13.585	{ 2.048	8.192	{ 4.096	12.288	4.096*	12.288	{ 8.192*	16.384
13.585–14	2.048	10.240	4.096	16.384	2.550	14.838	8.192	24.576
14–15	10.240	20.480	20.480	36.864	6.144	20.982	40.960	65.536
15–20	10.240	30.720	20.480	57.344	30.720	51.702	40.960	106.496
20–25	10.240	40.960	20.480	77.824	30.720	82.422	40.960	147.456
25–30	20.480	61.440	40.960	118.784	30.720	112.142	81.920	229.376
30–40	40.960	102.400	81.920	200.704	61.440	174.582	163.840	393.216
40–60					122.880	297.462		

[a] Growth is assumed exponential until the vitamin exerts its limiting effect. Thereafter, arithmetic growth is assumed at a rate proportional to vitamin concentration. Calculated values of cell growth in each time interval (CGTI) and cumulative cell count (CCC) are tabulated. The inoculum n is set at 10,000. *, End of exponential growth.

6.3.4 Vitamins as Limiting Factors: Practice

The conclusions reached from consideration of a mathematical model in Section 6.3.3 give only a qualitative idea of how the dose–response line may be expected to vary according to assay conditions. The value of the conclusions is that the analyst is provided with a logical basis for the introduction of modifications to established or recommended methods with a view to approaching the ideal of a rectilinear relationship between dose and response. It is necessary to determine by experiment to what extent it is practicable to increase incubation time and to what extent dose levels may be reduced without making the observed optical response so small that errors in its measurement become unacceptably large. Results

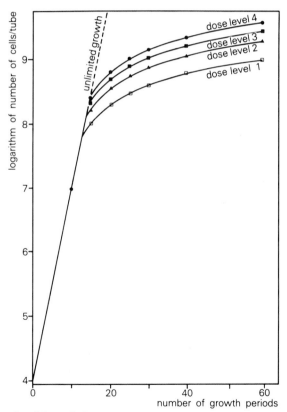

Fig. 6.6. A series of theoretical curves representing the growth of a test organism according to differing limiting doses of an essential vitamin. This is based on the same data as Fig. 6.5 but growth is shown on a logarithmic scale.

Table 6.4

Cumulative Cell Counts after Increasing Incubation Periods When Growth Rate Is Limited by Differing Dose Levels of One Essential Vitamin[a]

Number of growth periods	Dose level 1		Dose level 2		Dose level 3		Dose level 4	
	$CCC \times 10^{-7}$	Ratio	$CCC \times 10^{-7}$	Ratio	$CCC \times 10^{-7}$	Ratio	$CCC \times 10^{-}$	Ratio
15	10.24	1.000	16.38	1.600	20.98	2.049	24.58	2.400
20	20.48	1.000	36.86	1.800	50.70	2.476	65.54	3.200
25	30.72	1.000	57.34	1.867	82.42	2.683	106.50	3.467
30	40.96	1.000	77.82	1.900	112.14	2.738	147.46	3.600
40	61.44	1.000	118.78	1.933	174.58	2.841	229.38	3.733
60	102.40	1.000	200.70	1.960	297.46	2.905	393.22	3.840

[a] The ratios of cumulative cell counts (CCC) at different dose levels are shown for each number of growth periods.

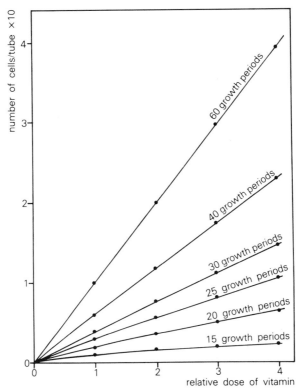

Fig. 6.7. An illustration of the predicted form of dose–response lines in a turbidimetric vitamin assay. These lines indicate that curvature will diminish as incubation time is increased. The lines are drawn from the same data as were used in Figs. 6.5 and 6.6.

published by many workers indicate that while some assays approach the ideal, in others the dose–response line is strongly curved. Some idea of the form of dose–response lines achieved in practice may be gained from the graphs of Fig. 6.8 for folic acid and vitamin B_{12}. Doses are shown on a single arbitrary scale for convenience of presentation. It should not be inferred that the same dose range is applicable to both assays.

6.3.5   Acid Production in Assays: Theory

The relationship between dose and observed response is more complex in the case of titrimetric than for turbidimetric assays for a variety of reasons. In general, product formation may coincide with cell growth, may occur after completion of cell growth, or may follow a pattern intermediate between these two extremes. Regardless of which pattern is followed, the

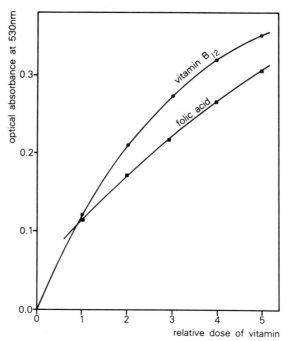

Fig. 6.8. Two practically determined dose–response lines in turbidimetric vitamin assays. The differences in curvature and in response to zero dose are indicative of the varying responses that may be encountered in this type of assay.

quantity of acid that is produced in an individual assay tube by the end of the incubation period is dependent *inter alia* on the number of cells and the time that these cells and/or the enzymes they contain have been in existence and able to form acid.

Separate mathematical models are suggested for amino acid and for vitamin assays.

6.3.5.1 Acid production in amino acid assays

The logic applied here to predict the rate of accumulation of acid is an extension of that of Section 6.3.1. The assumptions are as follows:

(i)   Growth is in accordance with the Monod equation [Eq. (6.1)], and so the number of cells per tube depending on dose level and incubation period is as in Table 6.1.

(ii)   Acid is produced at a rate that is constant for each cell throughout the incubation period. (It will be shown later that although this is not strictly true, little error results from the assumption.)

Let n equal the number of cells per tube at the time of inoculation, t, the "unit time period" (in minutes) which is identical with the generation time when all nutrients are abundant and growth is strictly exponential; z, the period of time (in minutes) during which cells produce acid, f, a factor relating number of cells producing acid to number of cells at time of inoculation, and a, the number of molecules of monobasic acid produced by n cells in t minutes. It follows that the number of molecules of monobasic acid A produced by fn cells in z minutes will be given by

$$A = afnz/t \qquad (6.2)$$

This simple equation is the basis of the calculation of predicted acid accumulation in an amino acid assay as illustrated by Table 6.5. Part A of Table 6.5 is based on a crude calculation which considers acid production over a period of $5t$ minutes in five discrete steps. Thus, the initial inoculum (not considering its growth) would in $5t$ minutes produce

$$a \times 1 \times n \times 5 = 5an \text{ molecules}$$

After t minutes the number of cells present (assuming dose level 1) would be $1.909ln$. The additional $0.909ln$ cells would in the following $4t$ minutes produce

$$a \times 0.9091 \times n \times 4 = 3.6364an \text{ molecules}$$

It is shown in part A of Table 6.5 that continuation of this calculation leads to a total of $26.784an$ molecules of acid produced. It is self-evident that if this calculation were carried out using 10 discrete steps to cover the five growth periods, the calculated total number of acid molecules produced would be somewhat greater. Part B of Table 6.5 shows how increasing the number of increments leads to higher values for the calculated number of molecules of acid produced. At 10 increments the value becomes $31.891an$ molecules and at 500 increments the calculated value is $37.525an$, which is approaching the limiting value. Naturally, a computer must be employed when the calculations involve so many increments.

The figures for acid accumulation that are given in Table 6.6 were calculated using increments of 0.01 of a unit growth period up to the point that cell multiplication ceased. Thereafter, acid production was calculated simply on the basis of the constant number of cells and the time of further incubation during which they produced acid at an assumed constant rate.

Figure 6.9 shows the form of dose–response curves that are predicted by the figures of Table 6.6 for acid accumulation. These indicate that the ideal of a linear response to dose would be approached only after a substantial number of growth periods.

Table 6.5

Method of Calculating Predicted Acid Accumulation in an Amino Acid
Assay[a]

(A) Acid production in five discrete steps

Time	RCCC	Increase in RCCC during previous growth period	Duration of acid production	Acid produced in period
0	1.0000	0.0000 – 1.0000[b]	5	5.0000
1	1.9091	1.9091 – 1.0000	4	3.6364
2	3.6443	3.6443 – 1.9091	3	5.2056
3	6.9562	6.9562 – 3.6443	2	6.6238
4	13.2750	13.2750 – 6.9562	1	6.3188
				26.7840

(B) Acid production increasing the number of increments

Size of increment	Predicted acid production in period
1.0	26.79
0.5	31.89
0.25	34.69
0.10	36.44
0.05	37.04
0.025	37.34
0.010	37.53
0.005	37.59
0.001	37.63
0.0001	37.65

[a] (A) Illustrates the principle of the calculation but, because it assumes
only five discrete steps to cover five growth periods, it seriously underestimates the acid produced. (B) shows how, by assuming a greater number of
smaller increments, the estimated figure for accumulated acid approaches a
limiting value.
[b] The inoculum.

6.3.5.2   Acid production in vitamin assays

The model used here differs from the previous model essentially in that
(i) growth is assumed to follow a strictly logarithmic rate as distinct from the
Monod model until the vitamin exerts its rate-controlling influence and the
arithmetic phase of growth begins. As in the case of acid production in
amino acid assays, the basis of the calculation of predicted acid accumulation is illustrated in tabular form. Part A of Table 6.7 is based on a crude

Table 6.6

Accumulation of Acid in an Amino Acid Assay[a]

Time units	Relative acid accumulation		
	Dose level 1	Dose level 2	Dose level 3
2	4.08	4.2	4.2
4	18.9	20.2	20.6
6	73.0	81.0	84.0
8	268.9	312.5	329.2
10	964.9	1,188.1	1,275.4
11	1,765.5	2,297.1	2,500.0
11.2	1,966.5	2,615.2	2,858.1
11.4	2,170.8	2,972.1	3,266.2
11.6	2,375.7	2,024.7	3,729.9
11.7	2,478.2[b]	—	—
11.8	—	3,772.4	4,254.4
11.9	—	3,977.3[b]	—
12.0	—	—	4,838.6
12.2	—	—	5,451.9[b]
13	3,811	6,231	7,910
14	4,836	8,280	10,983
16	6,886	12,378	17,129
18	8,936	16,476	23,275
20	10,986	20,574	29,421
25	16,111	30,819	44,786
30	21,236	41,064	60,151
35	26,361	51,309	75,516
40	31,486	61,554	90,881
50	41,736	82,044	121,611
60	51,986	102,534	152,341
100	92,986	184,494	275,261
150	144,236	286,944	428,911
200	195,486	389,394	582,561
240	236,486	471,354	705,481

[a] This assumes cell growth in accordance with the Monod equation [Eq. (6.1)] and a constant rate of acid production per cell. The unit of time is equal to the generation time when all nutrients are abundant.

[b] These figures coincide with the attainment of maximum cell count.

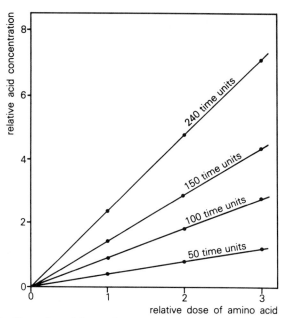

Fig. 6.9. An illustration of the predicted form of the dose–response lines in a titrimetric amino acid assay. The values that are plotted are taken from Table 6.6. These lines indicate that diminished curvature may be expected as incubation time is increased.

calculation which considers acid production in five discrete steps. Part B of Table 6.7 shows how more realistic figures are obtained as the number of discrete steps is increased to approach a limiting value.

To obtain the figures that are given in Table 6.8, representing predicted acid accumulation at various time intervals for three dose levels, time increments of 0.01 growth period were used. The form of dose–response curves that are predicted from these values is shown in Fig. 6.10. It is concluded, as in the model for amino acid assays, that the ideal of linearity is approached only after a substantial period of incubation.

6.3.6 Acid Production in Assays: Practice

It was predicted from the models of Section 6.3.5 that even assuming a constant rate of acid production by each cell, linearity of the dose–response curve would be approached only after many growth periods. This would be the expected situation in both amino acid and vitamin assays. In practice, the rate does not remain constant but decreases with time. Longsworth and MacInnes (1936) studied the growth and rate of acid production of *Lacto-*

Table 6.7

Method of Calculating Predicted Acid Accumulation in a Vitamin Assay[a]

(A) Acid accumulation in five discrete steps

Time	RCCC	Increase in RCCC during previous growth period	Duration of acid production	Acid produced in period
0	1	$1 - 0 = 1^b$	5	5
1	2	$2 - 1 = 1$	4	4
2	4	$4 - 2 = 2$	3	6
3	8	$8 - 4 = 4$	2	8
4	16	$16 - 8 = 8$	1	$\underline{8}$
				31

(B) Acid accumulation increasing the number of increments

Size of increment	Predicted acid production in period
1.0	31.00
0.5	37.42
0.25	40.96
0.10	43.19
0.05	43.95
0.025	44.34
0.010	44.57
0.005	44.65
0.001	44.71

[a] This is analogous to Table 6.5. (A) illustrates the principle of the calculation but, because it assumes only five discrete steps to cover five growth periods, it seriously underestimates the acid produced. (B) shows how, by assuming a greater number of smaller increments, the estimated figure for accumulated acid approaches a limiting value.

[b] The inoculum.

bacillus acidophilus under various conditions. Their work was not undertaken with reference to development of assay methods and involved liter-scale cultures from which aliquots were taken at intervals to measure viable cell counts and acid production. The result of one such experiment which is very relevant to assay work is summarized graphically.

In Fig. 6.11, the change in cell count and acid accumulation with time, over a period of 4 days is shown, both cell count and acid production being plotted on a log scale. Accumulation of acid is also shown on an arithmetic scale in Fig. 6.12. From these same data Longsworth and MacInnes derived

Table 6.8

Accumulation of Acid in a Vitamin Assay[a]

Time units	Relative acid accumulation		
	Dose level 1	Dose level 2	Dose level 3
1	1.4	1.4	1.4
2	4.3	4.3	4.3
4	21.6	21.6	21.6
6	90.6	90.6	90.6
8	366.6	366.6	366.6
10	1,471*	1,471	1,471
11	2,751	2,943*	2,943
11.585[b]			4,441*
12	4,543	5,503	5,848
15	12,991	19,327	23,887
20	37,311	62,847	84,672
30	0.124×10^6	0.227×10^6	$0,321 \times 10^6$
40	0.263×10^6	0.493×10^6	0.712×10^6
60	0.693×10^6	1.333×10^6	$1,953 \times 10^6$
100	2.167×10^6	4.241×10^6	6.280×10^6
150	5.162×10^6	10.180×10^6	15.143×10^6
200	9.438×10^6	18.679×10^6	27.847×10^6
240	13.780×10^6	27.322×10^6	40.775×10^6

[a] This assumes exponential growth up to the level at which the vitamin becomes a limiting factor (marked *). Thereafter growth is assumed to continue at a rate which is constant for each dose level and is equal to the final rate attained during exponential growth. A constant rate of acid production per cell is assumed throughout the period. The unit of time is equal to the generation time of the period of exponential growth.

[b] In the case of dose level 3, exponential growth ends after 11.585 generations.

figures for rate of acid production for a single average cell. The change in this parameter with time is shown in Fig. 6.13 using an arithmetic scale. From Fig. 6.13 it may be seen that rate of acid production was highest in the early part of the growth period. A value of about 26×10^{-15} mol/cell/hr was recorded at about 10 hours but dropped to about 4×10^{-15} mol/cell/hr by about 50 hours. It is of interest to note that these figures correspond to about 1.5×10^{10} and 0.25×10^{10} molecules/cell/hr, respectively. The rate of production of acid before 10 hours was not stated; it may be presumed

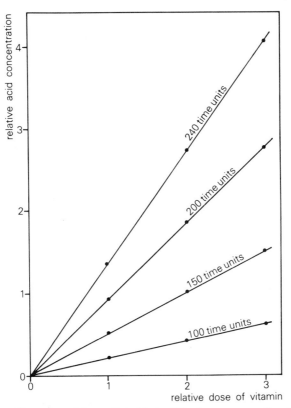

Fig. 6.10. An illustration of the predicted form of the dose–response lines in a titrimetric vitamin assay. The values that are plotted are taken from Table 6.7. As in the case of Fig. 6.9, these lines indicate that diminished curvature may be expected as incubation time is increased.

that the rate during this period was also high but that the number of cells present to produce acid was so low that the quantity of acid was too small to be measured. If it can be assumed that reduction in rate of acid production per cell with time is typical of organisms used in titrimetric assays, then this would be one of the factors contributing to nonlinearity of dose–response lines. However, by far the greater part of acid production is in the latter part of the incubation period when the rate of production is leveling out. Thus, the assumption of a constant rate of acid production in Section 6.3.5.1 was not unreasonable.

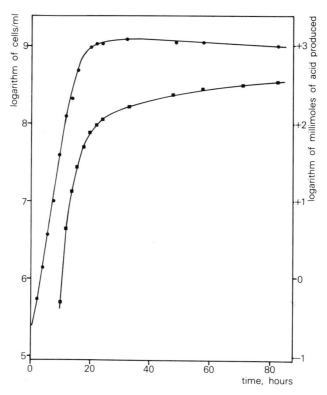

Fig. 6.11.   These graphs show the change in cell count and the accumulation of acid, both on logarithmic scales, during the growth of *Lactobacillus acidophilus.* From the work of Longsworth and MacInnes (1936). ●, Change in cell count; ■, accumulation of acid.

Some actual assay dose–response lines are presented in Fig. 6.14 which illustrate the differing attainment of approach to the ideal for a variety of assays and conditions. As in the case of Fig. 6.8, doses have been shown on a single arbitrary scale and it should not be inferred that the same dose range applies to all assays shown in the graph.

6.3.7   Comparison of Turbidimetric and Titrimetric Responses

Assay methods are often described which, with some slight modification, may be adapted to either turbidimetric or titrimetric response measurement. These modifications are usually to dose of growth-promoting substance being assayed and to duration of incubation. For the reasons that follow, the duration of incubation period is the main feature in which there are differences between the two variations of the basic method. Suppose the

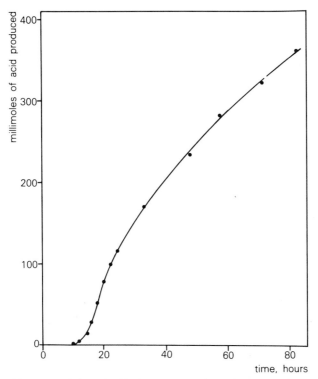

Fig. 6.12.   The accumulation of acid, shown on an arithmetic scale, during the growth of *Lactobacillus acidophilus*. This is based on the same data as the corresponding curve of Fig. 6.11, taken from the work of Longsworth and MacInnes (1936).

inoculum were of 10^4 cells/tube and that growth, as measured by turbidity, were not discernible until the level had reached 10^7 cells/tube. Then the increase of 1000-fold would correspond to about 10 generations ($2^{10} =$ 1024). If generation time were 1 hour or less, then an overnight incubation of around 16 hours would provide very adequate time for growth to give a conveniently measurable turbidimetric response. However, it is necessary that the incubation period be sufficient to permit maximum growth or at least the stationary phase of growth to be attained in tubes of all dose levels. Vincent (1985) reports that whereas overnight incubation may give sufficient turbidity for an optical measurement to be made, it is often necessary to continue incubation to a total of about 20 hours to allow sufficient growth in the higher-level tubes. Cell concentrations might be about 10^8– 10^9 per tube in the higher-dose level tubes at the end of the required incubation period in a turbidimetric assay. Suppose now that in a titrimet-

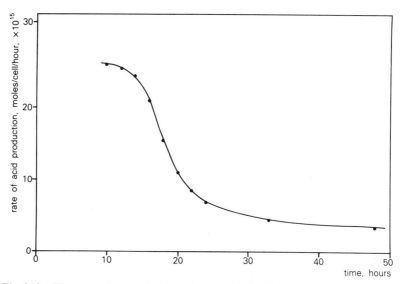

Fig. 6.13. The change in rate of acid production with time for an average single cell during the growth of *Lactobacillus acidophilus*. From the work of Longsworth and MacInnes (1936).

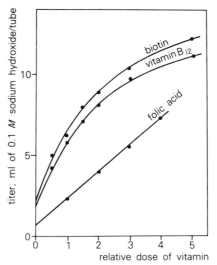

Fig. 6.14. Some practically determined dose–response lines in titrimetric assays of vitamins. Note that the dose scale shows *relative* doses for each vitamin; actual doses of the different vitamins are not compared.

ric assay, the dose of growth-promoting substance is such that cell concentrations in the higher-dose level tubes reach a maximum (assuming an amino acid assay) of about 10^{10} cells/tube; then consider the likely acid-producing capacity of such a cell suspension. As stated in Section 6.3.6, the work of Longsworth and MacInnes indicated that the rate of acid production leveled off down to 0.25×10^{10} molecules/cell/hr. If this is a typical value, then acid production in a tube containing 10^{10} cells/tube would be 0.25×10^{20} molecules/hr. As 1 mol contains about 6×10^{23} molecules, this rate corresponds to about 0.042 mmol/hr or 1 mmol of acid in 24 hours. It follows that normally an incubation period well in excess of 24 hours will be needed to attain a sufficiently large titrimetric response. Hence incubation periods of 48–72 hours are frequently specified for titrimetric assays.

6.4 Critical Factors

Among the most critical factors in assays for growth-promoting substances are cleanliness of all glassware used in the assay and the accurate preparation of test solutions at the very low concentrations that are used in some assays. These factors have been mentioned already in Sections 3.4.2 and 3.4.3 of Chapter 3, which describe the specific problems pertaining to microbiological assay.

Because the incubation period is longer than in the case of assays for growth-inhibiting substances and continues beyond the log phase of growth, the influences of initial inoculum size and exact duration of incubation are not necessarily so critical. However, for the same reason, contamination by extraneous microorganisms could be a problem. Consider separately the two cases of change from logarithmic to stationary phase and from logarithmic to arithmetic phase.

(1) *Change from logarithmic to stationary phase:* Assuming that the change is from log growth to an absolutely stationary phase, and that incubation time is adequate to ensure that all tubes in the assay (regardless of inoculum size and variations in incubation temperature) have reached the stationary phase, then variations in size of inoculum, temperature, and exact duration of incubation will be without influence on the final cell population.

(2) *Change from logarithmic to arithmetic phase:* In this case the influence of both initial inoculum and rate of growth, as might be influenced by minor differences in temperature, must be considered.

(a) Suppose that temperature of incubation is the same and is constant for all tubes but that the size of the inoculum varies as a result of the standard practice of adding one drop (as distinct from a precisely measured volume) of cell suspension to each tube. If the cell counts in two tubes were in the ratio 100:85, then the pattern of increase in cell count in the two tubes may be predicted subject to making certain assumptions: (i) inoculum in tube 1 is n cells, (ii) inoculum in tube 2 is $0.85n$ cells, and (iii) exponential growth changes abruptly to arithmetic growth when the level of $1024n$ cells/tube has been attained (this corresponds to 10 generations in tube 1 but to 10.23 generations in tube 2). The rate of increase in cell count at any instant during exponential growth is given by $\ln 2 \times$ cell count per growth period. Thus, at the time when cell count has reached $1024n$ cells/tube, the rate of increase will be $0.693 \times 1024n = 710n$ cells/growth period. Growth would then proceed in the arithmetic phase at a constant rate of $710n$ cells/growth period.

Calculations on the basis of these assumptions led to the values of cell counts for various numbers of growth periods that are presented in Table 6.9. It will be noted that the ratio of cell counts between the two tubes approaches 1.000 as incubation time is increased.

(b) Suppose that the inoculum level in all tubes is identical but that due to minor variations in temperature at different locations in the incubator bath there are slight differences in growth rate. As in the first example, the influence of such minor variations on the pattern of increase in cell count in two tubes may be predicted by making certain assumptions. In this case the assumptions are (i) inoculum in both tubes is n cells, (ii) generation time in

Table 6.9

Cell Counts for Various Numbers of Growth Periods at Constant Temperature Starting from Different Cell Concentrations

Number of growth periods	Tube 1 cell concentration	Tube 2 cell concentration	Cell concentration ratio (tube 1/tube 2)
0	n	$0.85n$	0.850
10	$1024n^a$	$870n$	0.850
10.23	$1190n$	$1024n^a$	0.861
11	$1734n$	$1571n$	0.906
12	$2444n$	$2281n$	0.933
13	$3154n$	$2991n$	0.948
14	$3864n$	$3701n$	0.958
15	$4574n$	$4411n$	0.964
20	$8124n$	$7961n$	0.980

[a] Indicates the end of exponential growth.

tube 2 is 1% greater than in tube 1, (iii) exponential growth changes abruptly to arithmetic growth in both tubes when the level of $1024n$ cells/tube has been attained, i.e., after 10 generations, (iv) unit "growth period" is equal to the generation time during exponential growth for tube 1; it follows that for tube 2 the generation time is 1.01 growth periods; (v) arithmetic growth in each tube continues at the same rate as the final exponential rate.

Again, as in the first example, the rate of growth in tube 1 will be $710n$ cells/growth period at the instant that the cell count has reached $1024n$. The rate of growth in tube 2 is related to that in tube 1 at all times by the factor $2^{100/101} - 1 = 0.9863$. Thus, whereas the number of cells in tube 1 after t growth periods during the log phase is given by $n \times 2^t$, that for tube 2 is given by $n \times 1.9863^t$.

The rate of growth in tube 2 at the instant that cell count reaches $1024n$ will be

$$0.9863 \times \ln 2 \times 1024n = 700n \text{ cells/growth period}$$

Calculations on the basis of these assumptions led to the values of cell counts for the various numbers of growth periods that are presented in Table 6.10. As in the case of Table 6.9, the ratio of cell counts between the two tubes approaches 1.00 as incubation time is increased.

It is concluded that although biases due to uneven inoculation and uneven incubation temperature tend to become less serious with increasing incubation time, they are best avoided through good techniques.

Table 6.10

Cell Counts for Various Numbers of Growth Periods at Slightly Different Temperatures

Number of growth periods	Tube 1 cell concentration	Tube 2 cell concentration	Cell concentration ratio (tube 1/tube 2)
0	n	n	1.000
10	$1024n^a$	$956n$	0.934
10.1	$1095n$	$1024n^a$	0.935
11	$1734n$	$1654n$	0.954
12	$2444n$	$2354n$	0.963
13	$3154n$	$3054n$	0.968
14	$3864n$	$3754n$	0.972
15	$4574n$	$4454n$	0.974
20	$8124n$	$7954n$	0.979

[a] Indicates the end of exponential growth.

6.5 The Tube Method in Practice

The essential steps in the assay are represented as a flowchart in Fig. 6.15. Maintenance of stock cultures and preparation of the inoculum have been described in Chapter 2. This work is necessarily done in advance of the day of the assay. The period of incubation varies considerably in different growth-promoting substance assays. For example, the Association of Official Agricultural Chemists (AOAC) in the United States has established two alternative assay methods for folic acid using *Streptococcus faecalis:* a titrimetric and a turbidimetric assay. For the titrimetric assay an incubation time of 72 hours is specified; for the turbidimetric assay (which uses dose

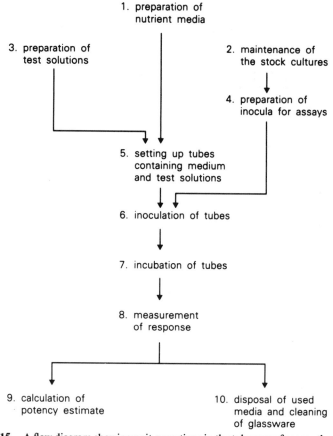

Fig. 6.15.   A flow diagram showing unit operations in the tube assay for growth-promoting substances.

levels of one-half of those in the titrimetric assay), an incubation period of only 16–24 hours is suggested. Another turbidimetric method using *Lactobacillus casei* (a slow-growing organism) requires about 40 hours of incubation. Clearly the incubation period needed must have a great bearing on general work planning in the laboratory.

6.5.1   Preparation of Assay Tubes

Formulas for media are often complex, as they are prepared from vitamin-free components to which are then added specific quantities of essential nutrients other than the growth-promoting substance that is to be assayed. Fortunately, these may be purchased from commercial sources as dry granules ready for reconstitution by the addition of water. A convenient routine for the preparation of tubes for an assay is first to make up a suitable quantity of liquid medium at double the recommended assay strength. Prepare test solutions of both reference standard and sample(s) at concentrations recommended for the assay of a particular substance. From these reference solutions, pipette, for example, 1, 2, and 3 ml of each solution into replicate tubes. In each case add sufficient water to make up to 5 ml. Additionally, set up two series of replicate tubes containing only 5 ml of water. Then to every tube add 5 ml of the double-strength assay medium. The arrangement of the tubes may be as shown in Fig. 6.16, for which the replication is three. The two sets of tubes containing only water and double-strength medium are to serve as inoculated and uninoculated blanks. Additionally, two or more tubes of the highest dose level of standard are desirable to check the progress of growth and determine the appropriate time to end incubation.

6.5.2   Heat Treatment of Tubes

All tubes must be identifiable with respect to preparation and dose level. This is preferably done by a code mark, durable for the period of the assay, made near the top of the tube or, alternatively, on its metal cap. After marking, the tubes are heat-treated by autoclaving or steaming as specified in the written assay procedure. This treatment is often referred to as "sterilization." Its purpose is to destroy living microorganisms that may contaminate the medium, but the treatment recommended is generally too mild to achieve sterilization. The conditions are a compromise aimed to minimize contamination with minimal destruction of essential nutrients. To avoid variations in the heat treatment of individual tubes that could lead to variable diminution of nutrients, tubes should be of uniform thickness and shape and should not be too closely packed during the heat treatment.

	Uninoculated blank	Inoculated blank	Reference standards				Sample of unknown potency		
Dose and replication	0*	0**	1	2	3	3***	1	2	3
	0*	0**	1	2	3	3***	1	2	3
	0*	0**	1	2	3		1	2	3

*   The uninoculated blank serves as the zero growth reference point for optical measurements, i.e. 100% transmittance in a spectrophotometer or zero response in a nephelometer.

**   The inoculated blank provides the response to zero dose which may be used in plotting a graph or in the purely arithmetical calculation of potency estimate that is described in section 6.5.5.

*** The additional high dose levels of reference standard are used solely to determine when growth has ceased; they are not used in the calculation of potency estimate.

Fig. 6.16. A chart showing a typical plan of the test for a three-dose level assay of a growth-promoting substance. This dose not correspond to the final layout of tubes in a rack, because their positions should be randomized for incubation.

6.5.3   Inoculation and Incubation

The tubes are cooled to about the intended incubation temperature; then, with the exception of the "uninoculated blanks," one drop of the previously prepared suspension of the inoculum is added to each and the cap replaced. The racks of tubes are placed in a well-stirred water bath controlled at $\pm 0.5\,^\circ$C of the desired incubation temperature. The actual variation within the bath should be within $\pm\,0.1\,^\circ$C of the mean. In the case of turbidimetric assays, when the required nominal incubation period has been almost completed, the turbidity of one of the extra tubes corresponding to the highest dose level is measured. The tube is replaced in the incubation bath and left to incubate for an additional period of perhaps 30 to 60 minutes. When there is no apparent increase in turbidity between two successive readings, the incubation may be stopped.

In the case of assays with a titrimetric measurement of response, the total incubation period is greater, and ± 1 hour in termination of the incubation does not make a great difference to the overall level of titrations. Termination of incubation then titration of tubes or flasks can be at a time that is most appropriate for the smooth completion of the day's work.

6.5.4   Measurement of Response

If the response to be observed is the increase in turbidity of the suspension arising from the multiplication of cell numbers, then all the facts that were considered in Chapter 5 with respect to assays of growth-inhibiting substances are equally applicable here. The reader is therefore referred to Section 5.5.3 of that chapter.

If optical properties of the suspension are measured with a spectropho-tometer or other absorptiometer, then absorbance should be recorded and used in the graphic or arithmetic estimation of sample potency. Even though the instruments actually measure the light that is transmitted by the cell suspension, it would be quite illogical to record and use transmittance directly in the potency estimation, because a plot of transmittance versus dose will not even approximate to a straight line.

If the response to be observed is the quantity of acid produced in individual tubes, then titrations should be carried out with the routine skills of analytical chemistry. A burette of 10 or 15 ml capacity, calibrated in divisions of 0.05 ml, would normally be appropriate.

6.5.5   Calculation of the Potency Estimate (Part 1)

In order that data derived from an experimental procedure may be processed mathematically as distinct from graphically, it is highly desirable that some measure of response, or a function of that response, be directly responsible to dose, or to a function of dose. The advantage of a mathemati-cal procedure is that it eliminates the errors of drawing the line of best fit to a series of points. Thus it can lead to the best estimate of potency that it is possible to derive from the experimental data. A statistical evaluation also becomes possible, so that fiducial limits of the potency estimates may be calculated.

A disadvantage of the purely mathematical approach is that any abnor-mality of the response line is not so easily visualized as in the graphic method.

The principles of the method of potency estimation for the type of assay under consideration may be seen readily from the ideal case, which is illustrated graphically in Fig. 6.17, and also from the more common case illustrated in Fig. 6.18, in which there is a small response in the zero-dose control tubes.

For the purpose of illustrating this calculation procedure the following postulations are made:

(i)   In the zero-dose control tubes there is a small response a, which is due to an adventitious trace of the growth-promoting substance.

(ii)   In all other individual tubes the response may be represented as $a + bx$, in which the two components are (a) the response a to the ad-ventitious trace of growth-promoting substance (which is assumed to be the same in all tubes including those for zero-dose control); and (b) a response bx in which b is the slope of the response line to an individual preparation and is directly proportional to its potency; and x, which is the dose level.

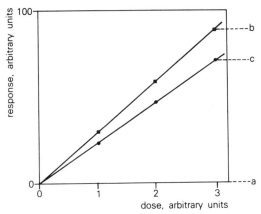

Fig. 6.17. An ideal response in an assay for a growth-promoting substance. This represents an assay in which there are three dose levels of both reference standard and a sample of unknown potency. Doses are in the ratio 1 : 2 : 3. This design is called a "seven-point common zero assay." The ratio *ac* : *ab* in this particular assay is 0.8, indicating that the sample has a potency of 80% of that of the reference standard. ■, Reference standard; ●, sample of unknown potency.

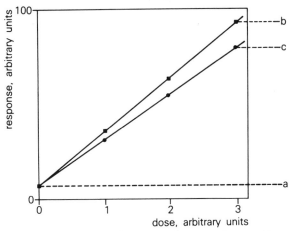

Fig. 6.18. An almost ideal response in an assay for a growth-promoting substance. This differs from the graph of Fig. 6.17 in that there is a small positive response at zero dose. The conditions for a valid slope ratio assay are still met because the two response lines are straight and coincide at zero dose. In this case they also coincide at zero dose with the actual response to the zero-dose control. However, this is not an essential condition because the response to zero control may be omitted from the calculation. The ratio *ac* : *ab* is 0.84 in this example, indicating that the sample has a potency of 84% of that of the reference standard. ■, Reference standard; ●, sample of unknown potency.

It follows that the ratio of the slopes of the two response lines should be the same as the ratio of the potencies of the test solutions which caused the responses. In both Fig. 6.17 and 6.18 the ratio of the slopes is ac/ab.

In drawing a graph, assuming the points plotted for each preparation appeared to correspond to a straight line, we should most probably draw a line ensuring that it lay close to the response to the highest dose rather than draw it close to the responses to the intermediate at the expense of a substantial deviation from the high response. We should thus be intuitively placing greater reliance on the response to the higher dose, or "weighting" that response.

The purely mathematical procedure incorporates a "weighting" of higher responses in a very precise way. First, it effectively separates the component of response that is evident in the zero-dose control tubes, then it introduces weighting factors to place more emphasis on responses to higher doses. Responses are weighted in direct proportion to their corresponding doses.

Mathematical formulas adaptable to a wide range of assay designs were devised by Bliss (1952). These may be applied regardless of number of dose levels, number of samples to be compared with a standard, and the replication employed. The Bliss formulas are somewhat more complex than those now introduced; however, they will be illustrated for comparison in Section 6.5.6. The principles of the simpler calculations are illustrated here by an example consisting of one sample, one standard each at three dose levels, together with a common zero-dose control. In the first place, for simplicity, a replication of unity is assumed; i.e., it is assumed that the assay includes only seven tubes in all.

There are five steps in the calculation:

(i) Calculate the sum of all responses $S(y)$.

(ii) Calculate the weighted sums of responses for standard and sample individually, T_S and T_T, then their sum $S(T_i)$.

(iii) From $S(T_i)$ and $S(y)$, calculate a', the best estimate of a, the response due to an adventitious trace of growth-promoting substance.

(iv) Calculate the individual slopes of response lines for standard b_S and sample b_T, respectively.

(v) Calculate the slope ratio b_T/b_S which is equal to the potency ratio.

To derive expressions by which these five steps may be carried out, it is first necessary to set out the components of the responses in each of the seven tubes. This is done in Table 6.11.

The expressions for each of the five steps are then derived using the tabulated data thus:

(i) $$S(y) = [a' + (a' + b_S) + (a' + 2b_S) + (a' + 3b_S) + (a' + b_T) +$$
$$(a' + 2b_T) + (a' + 3b_T)]$$

which simplifies to

$$S(y) = 7a' + 6b_S + 6b_T \tag{6.3}$$

(ii)　　$T_S = [0(a') + 1(a' + b_S) + 2(a' + 2b_S) + 3(a' + 3b_S)]$

which simplifies to

$$T_S = 6a' + 14b_S \tag{6.4}$$

Similarly,

$$T_T = 6a' + 14b_T \tag{6.5}$$

and

$$S(T_i) = T_S + T_T = 12a' + 14b_S + 14b_T \tag{6.6}$$

(iii)　　a' is then obtained by the use of simultaneous equations. Multiply Eq. (6.3) by 7 and subtract from the product, Eq. (6.6) multiplied by 3:

$$7S(y) = 49a' + 42b_S + 42b_T$$

$$3S(T_i) = 36a' + 42b_S + 42b_T$$

so that

$$7S(y) - 3S(T_i) = 49a' - 36a' = 13a'$$

and

$$a' = \frac{7S(y) - 3S(T_i)}{13} \tag{6.7}$$

(iv)　　The slopes b_S and b_T may now be calculated by rearranging Eq. (6.3) and (6.4) thus:

Table 6.11

Ideal Response to Dose[a]

Preparation	Dose (x)	Response (y)
Zero control	0	a'
Standard	1	$a' + b_S$
	2	$a' + 2b_S$
	3	$a' + 3b_S$
Sample	1	$a' + b_T$
	2	$a' + 2b_T$
	3	$a' + 3b_T$

[a] From Hewitt (1977).

$$b_S = T_S - \frac{6a'}{14} \qquad\qquad (6.8)$$

$$b_T = T_T - \frac{6a'}{14} \qquad\qquad (6.9)$$

(v)   Calculate the potency ratio R as

$$R = b_T/b_S \qquad\qquad (6.10)$$

These formulas may be adapted readily for other numbers of dose level and for any degree of replication. They represent a particular case of the general equations of Bliss. The application of this type of formulas is illustrated in Example 6.1.

The data used to illustrate some calculation procedures are from a turbidimetric assay of nicotinic acid from Vincent (1978). The assay as actually carried out comprised a standard against which six samples were compared. From each of the seven preparations, test solutions at five dose levels were prepared; from each of the 35 treatments thus prepared, triplicate-assay tubes were set up, making 105 excluding blanks. These tubes were distributed in three racks.

The observed response was the percentage transmittance of monochromatic light in a simple spectrophotometer. Responses were recorded as percentage transmittance, and relative potencies were calculated directly by means of a computer program that fitted a mathematical model to the curved response lines.

The computer printout is reproduced in Fig. 6.19, and the curved plot of percentage transmittance against dose is shown in Fig. 6.20 for the reference standard and for samples A and B only. The computer calculation procedure is not described here as this is outside the intended scope of this book.

For the application of calculations based on the assumption of a slope ratio relationship between dose and response, the raw data must be changed from percentage transmittance to optical absorbance. A plot of the thus-transformed responses against dose, it is hoped, might approximate to a straight line. Transformed responses together with their means for each treatment are presented in Table 6.12 for the reference standard and for samples A and B only. The means of transformed responses are plotted against dose in Fig. 6.21, from which it is seen that the hoped-for linearization was not attained, although curvature was not strong.

In example 6.1, which follows, only a portion of the modified data is used to illustrate the simple seven-point assay in which only one sample is compared with the standard. For this purpose the three lower dose levels of

29/06/78   TUBE ASSAY

NICO

STANDARD   % Transmittance at Dose Level

		1	2	3	4	5
OBS'D		80.0	67.0	59.0	51.0	44.0
		82.0	68.0	58.0	51.0	44.0
		80.0	69.0	59.0	51.0	45.0
	MEAN	80.7	68.0	58.7	51.0	44.3
EST'D		80.9	68.2	58.3	50.7	44.7

SAMPLE   POTENCY   PC   S.E.

1	1.92*E 03	0.6

OBS'D		82.0	69.0	60.0	52.0	45.0
		84.0	71.0	61.0	52.0	46.0
		82.0	70.0	60.0	54.0	47.0
	MEAN	82.7	70.0	60.3	52.7	46.0
EST'D		81.9	69.8	60.2	52.7	46.7

2	1.96*E 03	0.6

OBS'D		82.0	68.0	59.0	51.0	47.0
		82.0	70.0	59.0	53.0	46.0
		81.0	69.0	60.0	53.0	46.0
	MEAN	81.7	69.0	59.3	52.3	46.3
EST'D		81.6	69.3	59.7	52.1	46.1

3	1.93*E 03	0.6

OBS'D		82.0	69.0	59.0	53.0	46.0
		81.0	71.0	59.0	54.0	46.0
		81.0	69.0	60.0	54.0	46.0
	MEAN	81.3	69.7	59.3	53.7	46.0
EST'D		81.8	69.6	60.0	52.4	46.4

Fig. 6.19. A computer printout of the observed responses in an assay of nicotinic acid together with calculation of the potency estimate and statistical evaluation. For each of the seven preparations (standard plus six samples), first the three responses to each treatment together with their means are printed. On the succeeding lines "EST'D" is the best estimate of the expected response based on all observed responses and the equations of their curves as calculated by the computer. A computer printout stating "1.92*E   03   06" means "potency 1920 μg/ml with a standard error of 0.6%." Observed responses that are recorded here are used in Examples 6.1, 6.2, and 6.3.

			% Transmittance at Dose Level				
			1	2	3	4	5
4	1.98*E	03	0.6				
	OBS'D		81.0	68.0	59.0	53.0	46.0
			82.0	68.0	60.0	52.0	45.0
			81.0	69.0	60.0	52.0	46.0
	MEAN		81.3	68.3	59.7	52.3	45.7
	EST'D		81.4	69.1	59.4	51.8	45.8
5	1.91*E	03	0.6				
	OBS'D		81.0	69.0	61.0	53.0	46.0
			82.0	70.0	60.0	53.0	47.0
			83.0	69.0	60.0	53.0	47.0
	MEAN		82.0	69.3	60.3	53.0	46.7
	EST'D		81.9	69.8	60.3	52.7	46.7
6	1.89*E	03	0.6				
	OBS'D		82.0	70.0	60.0	53.0	47.0
			82.0	71.0	60.0	53.0	46.0
			83.0	70.0	60.0	53.0	48.0
	MEAN		82.3	70.3	60.0	53.0	47.0
	EST'D		82.0	70.0	60.5	53.0	47.0

SOURCE	DF	MEAN SQR
FITTED MODEL	9	1850.23
DEV FROM MODEL	14	0.86
DEV FROM SIM	12	0.51
RACKS	2	2.30
ERROR	68	0.49

TUBE S.E.   =   0.7

Fig. 6.19   (*Continued*)

standard and sample A are used. Because there are no responses to a zero dose recorded in this example, a modified expression corresponding to Eq. (6.7) is derived in which the denominator is 6 instead of 13.

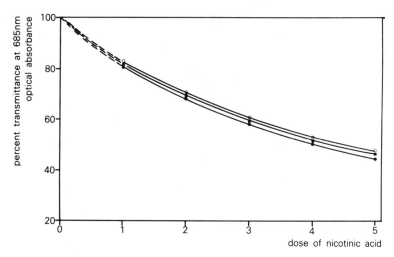

Fig. 6.20. Dose–response curves from an assay of nicotinic acid. The means of observed responses (percentage transmittance) that were recorded in the computer printout of Fig. 6.19 are plotted against relative dose of nicotinic acid. Data for the reference standard and first two samples only are presented. ●, Reference standard; ○, first sample; ■, second sample.

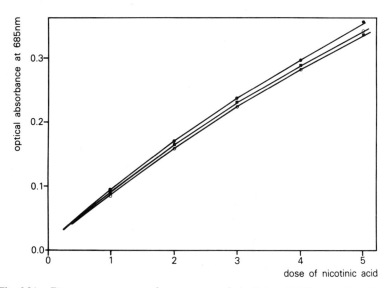

Fig. 6.21. Dose–response curves form an assay of nicotinic acid. These are based on the same observed data as the curves of Fig. 6.20, but means of percentage transmittance have been converted to optical absorbance, A. Optical absorbance is calculated from percentage transmittance T, by the expression $A = \log 100 - \log T$. ●, Reference standard; ○, first sample; ■, second sample.

6.5.6   Calculation of the Potency Estimate (Part 2)

The whole of the data presented in Table 6.12 will be used in Example 6.2 to calculate potency estimates. The expressions of Bliss (1952) for a' and b are used. These expressions, presented without explanation of their derivation, are as follows:

$$a' = \frac{2(2k+1)S(y) - 6S(T_i)}{N(k-1) + 3h'(k+1)} \qquad (6.11)$$

$$b_i = [3/(2k+1)]\{[2T_i/fk(k+1)] - a'\} \qquad (6.12)$$

in which f is the degree of replication at each dose level of all preparations, h', degree of replication at zero dose, k, number of dose levels (excluding zero), and N, total number of observed responses (including zero).

Example 6.1:   Assay of Nicotinic acid, Three Dose Levels

Test organism: Lactobacillus plantarum NCIB 8030, ATCC 8014
Experimental design: seven-point common zero
Weighings and dilutions:
  Reference standard

$$103.1 \text{ mg} \rightarrow 1000 \text{ ml}: 5 \text{ ml} \rightarrow 500 \text{ ml}: 5 \text{ ml} \rightarrow 250 \text{ ml}$$

  "Unknown" sample — multivitamin syrup*

$$1.344 \text{ g} \rightarrow 50 \text{ ml}: 5 \text{ ml} \rightarrow 500 \text{ ml}: 5 \text{ ml} \rightarrow 100 \text{ ml}$$

Note: 1 ml is 1.344 g.

Responses, sums of three absorbances for each of dose levels 1, 2, and 3 for standard and sample A, from Table 6.12

	Response		
Preparation	Dose 1	Dose 2	Dose 3
Standard	0.280	0.502	0.695
Sample A	0.248	0.465	0.659

Table 6.12

Turbidimetric Assay of Nicotinic Acid[a]

	Dose level					Sum of absorbances
Preparation	1	2	3	4	5	
Standard	0.097	0.174	0.229	0.292	0.357	3.414
	0.086	0.167	0.237	0.292	0.357	
	0.097	0.161	0.229	0.292	0.347	
Total	0.280	0.502	0.695	0.876	1.061	
Mean absorbance	0.093	0.167	0.232	0.292	0.354	
Sample A	0.086	0.161	0.222	0.284	0.347	3.220
	0.076	0.149	0.215	0.284	0.337	
	0.086	0.155	0.222	0.268	0.328	
Total	0.248	0.465	0.659	0.836	1.012	
Mean absorbance	0.083	0.155	0.220	0.279	0.337	
Sample B	0.086	0.167	0.229	0.292	0.328	3.273
	0.086	0.155	0.229	0.276	0.337	
	0.092	0.161	0.222	0.276	0.337	
Total	0.264	0.483	0.680	0.844	1.002	
Mean absorbance	0.088	0.161	0.227	0.281	0.334	

[a] Observed responses that were initially recorded as percentage transmittance at 530 nm have been transformed to optical absorbance.

Calculation:

$$S(y) = 0.280 + 0.502 + 0.695 + 0.248 + 0.465 + 0.659 = 2.849$$

$$T_S = 0.280 + 2(0.502) + 3(0.695) \qquad = 3.369$$

$$T_A = 0.248 + 2(0.465) + 3(0.659) \qquad = 3.155$$

$$S(T_i) = 3.369 + 3.155 \qquad = 6.524$$

From the assay design the parameters for the Bliss equation are

$$f = 3, \ h' = 0, \ k = 3, \ N = 18$$

Substituting these parameters in Eqs. (6.11) and (6.12) we obtain:

$$a' = \frac{(2 \times 7)(2.849) - 6(6.524)}{18(3-1)+0} = 0.02061$$

$$b_S = \frac{3}{7}\left\{\left[\frac{2 \times 3.369}{(3 \times 3)(3+1)}\right] - 0.02061\right\} = 0.07138$$

$$b_A = \frac{3}{7}\left\{\left[\frac{2 \times 3.155}{(3 \times 3)(3+1)}\right] - 0.02061\right\} = 0.06629$$

$$R = \frac{0.06629}{0.07138} = 0.9287$$

Standard test solution potency $= \dfrac{103.1 \times 5 \times 5}{1000 \times 500 \times 250}$ as mg/ml or 0.02062 μg/ml.

Therefore, sample potency $= \dfrac{0.02062 \times 0.9287 \times 100 \times 500 \times 50}{5 \times 5 \times 1.0}$ as μg/ml or 1.915 mg/ml.

Example 6.2:   Assay of Nicotinic Acid, Five Dose Levels

Test organism: L. plantarum NCIB 8030, ATCC 8014
Experimental design: Five-dose level slope ratio for two samples
Weighings and dilutions:
  Reference standard

$$103.1 \text{ mg} \rightarrow 1000 \text{ ml}:5 \text{ ml} \rightarrow 500 \text{ ml}:5 \text{ ml} \rightarrow 250 \text{ ml}$$

Sample A

$$1.344 \text{ g}^* \rightarrow 50 \text{ ml}:5 \text{ ml} \rightarrow 500 \text{ ml}:5 \text{ ml} \rightarrow 100 \text{ ml}.$$

Sample B

$$1.351 \text{ g}^{**} \rightarrow 50 \text{ ml}:5 \text{ ml} \rightarrow 500 \text{ ml}:5 \text{ ml} \rightarrow 100 \text{ ml}$$

Note: * 1 ml is 1.344 g.
        ** 1 ml is 1.351 g.
Responses, sums of three absorbances for each of five dose levels for standard and two samples, from Table 6.12

Preparation	Response					
	Dose 1	Dose 2	Dose 3	Dose 4	Dose 5	Totals
Standard	0.280	0.502	0.695	0.876	1.061	3.414
Sample A	0.248	0.465	0.659	0.836	1.012	3.220
Sample B	0.264	0.483	0.680	0.844	1.002	3.272
						$S(y) = 9.907$

Calculation:
From the response totals for each treatment calculate:

$$T_S = 0.280 + 2(0.502) + 3(0.695) + 4(0.876) + 5(1.061) = 12.178$$
$$T_A = 0.248 + 2(0.465) + 3(0.659) + 4(0.836) + 5(1.012) = 11.559$$
$$T_B = 0.264 + 2(0.483) + 3(0.680) + 4(0.844) + 5(1.002) = 11.656$$

then $S(T_i) = 12.178 + 11.559 + 11.656 = 35.393$.
From the assay design the parameters for the Bliss equation are

$$f = 3, h' = 0, k = 5, N = 45.$$

Substituting these parameters in Eqs. (6.11) and (6.12) we obtain:

$$a' = \frac{(2 \times 11)(9.907) - 6(35.393)}{45(5 - 1) + 0} = 0.03109$$

and

$$b_S = 3/11 \left\{ \left[\frac{2 \times 12.178}{(3 \times 5)(5 + 1)} \right] - 0.03109 \right\} = 0.06533$$

similarly, $b_A = 0.06158$ and $b_B = 0.06216$, so that the dose ratios corresponding to the slope ratios are, for sample A/standard,

$$\frac{0.06158}{0.06533} = 0.9426$$

and for sample B/standard,

$$\frac{0.06216}{0.06533} = 0.9515$$

Standard test solution potency $= \dfrac{103.1 \times 5 \times 5}{1000 \times 500 \times 250}$ as mg/ml or 0.02062 μg/ml.

Therefore, sample potencies for sample A

$$\frac{0.02062 \times 0.9426 \times 100 \times 500 \times 50}{5 \times 5 \times 1.0} \text{ as } \mu\text{g/ml, or 1.944 mg/ml}$$

for sample B

$$\frac{0.02062 \times 0.9515 \times 100 \times 500 \times 50}{5 \times 5 \times 1.0} \text{ as } \mu\text{g/ml, or 1.962 mg/ml}$$

6.5.7   Graphic Interpolation of Potency

Interpolation of sample potency from the standard dose–response line is an obvious and simple procedure. Commonly, test solutions may be prepared from the sample at two or three dilutions then their responses be used to interpolate from a standard dose–response line based on perhaps five dose levels. In Example 6.3, the data of Table 6.12 are used once again; the mean responses to the middle three dose levels of samples are used to interpolate sample potencies.

Example 6.3:   Assay of nicotinic acid, Graphic interpolation

Experimental details: Identical with Example 6.2
Interpolation and potency estimation:

Sample	Nominal potency	Interpolated potency	Potency ratio Unknown/standard	Mean	Estimated sample potency (mg/ml)
A	2	1.85	0.925		
	3	2.80	0.933	0.934	1.926
	4	3.77	0.943		
B	2	1.92	0.960		
	3	2.92	0.973	0.963	1.986
	4	3.82	0.955		

It is of interest to compare the estimated potencies obtained in these three examples and also that obtained by the computer program as shown in Fig. 6.19. These are summarized in Table 6.13.

Table 6.13

Potency Estimates Obtained from the Same Data in an Assay of Nicotinic Acid but Using Different Calculation Procedures

Calculation Procedure	Reference	Potency estimates Sample A	Sample B
Computer program incorporating data from standard and six samples	Computer printout, (Fig. 6.19)	1.920	1.960
Three-dose level assay, calculation for sample A only	Example 6.1	1.915	—
Five-dose level assay for two samples compared with one standard, Bliss calculation procedure	Example 6.2	1.944	1.962
Graphic interpolation	Example 6.3	1.926	1.986

References

Bliss, C. I. (1952). "The Statistics of Bioassay." Academic Press, New York.
Hewitt, W. (1977). "Microbiological Assay," p. 94. Academic Press, Orlando.
Longsworth, L. G., and MacInnes, D. A. (1936). *J. Bacterial* **31**, 292.
Monod, J. (1949). *Ann. Rev. Microbiol.* **3**, 371.
Toennies, G., and Shockman, G. D. (1953). *Arch. Biochem. Biophys.* **45**, 447.
Vincent, S. (1978). Unpublished work.
Vincent, S. (1985). Unpublished work.

CHAPTER 7

QUALITY CONTROL PROCEDURES

7.1 Introduction

It is possible to produce meaningful results and to achieve a precision similar to many chemical analytical methods by means of microbiological assay techniques and on a regular basis by the establishment of rigorous control methods that do not leave anything to chance. A dedicated assay leader takes nothing for granted and knows that the smooth running of a microbiological assay unit can be assured only by the continuous monitoring of all the controllable variables which can affect the accuracy and the precision of the results.

A surprisingly large number of control methods are necessary to ensure that all assistants perform their allotted duties to the best of their ability, that all the assay cultures are in prime condition, and that all media and equipment are functioning according to specification. Only then is it possible to obtain results which can be quoted with full confidence. There are assay laboratories where the importance of controlling external factors which can affect assay results is not fully realized and avoidable bias in the work goes unrecognized, resulting in the application of statistical calculations to biased or doubtful data and assay reports that suggest confidence which is entirely unwarranted.

The main parameters affecting microbiological assays are well documented (Lees and Tootill, 1955; Cooper, 1963 and 1972), but considerations of theory and mathematical calculations are usually based on ideal conditions with the assumption that all systematic errors have been eliminated. Kavanagh and Ragheb (1979) summarized some of the essential considerations necessary for assuring accuracy in the microbiological assays for vitamins and antibiotics. They point out very rightly that poor results are usually caused by correctable defects in operations or equipment.

Over the past 10-15 years the emphasis has been moving gradually to more and more precision in assays due to the demands of the research chemists who want to establish purity and stability of an antibiotic compound. They are looking for results which are in line with physical or chemical methods of assay with errors of 2% or less. Consequently, a conscious effort has been made in recent years to eliminate or at least greatly reduce the bias which can be associated with manual reading by introducing automated or semiautomated optical measuring equipment. Another notable improvement has been the widespread use of accurate pipetting aids which are fitted with nonwettable polypropylene tips delivering standard volumes of solutions instead of "filling cylinders to the top" or counting the drops being delivered. Automatic pickup of standard and sample solutions, media, and inoculum frequently used with turbidimetric assays are other examples. The introduction of computerized calculations of results and analysis of variance is yet another example whereby mistakes due to carelessness or fatigue can be eliminated. The adoption of sound analytical techniques, attention to detail, and the elimination of guesswork will soon pay dividends.

The following paragraphs are intended to show that with the adoption of regular monitoring throughout every stage of the assay it is possible to reduce human and instrumental errors and to approach the optimal assay conditions so necessary if a high-precision assay is to be achieved and maintained. Only this way will it be possible to destroy the myth of the "biological error."

7.2 Media

Good laboratory practice requirements demand that each bottle of medium be clearly identified to show the contents, the batch number, and the date of expiry. The use of indelible ink is permitted, but it is much neater to use stick-on labels which can be applied easily and quickly with a supermarket-type labeling gun. The label and markings must withstand the effect of steam during sterilization and subsequent steaming during remelting, as well as the effect of the hot water while the bottles are in the water bath cooling before being inoculated.

It is convenient to keep media in bottles with polypropylene screw caps, which can be obtained in a minimum of three standard colors: white, red, and black. This color coding can be used to identify three different types of assay media, thereby reducing the labeling requirements. It is quite easy to add other colors by painting over the white tops with marking ink, which is available in many different colors, such as blue, yellow, green, and mauve. It is also possible to use a combination of two different colors, for example, blue/white or red/yellow. If the white tops are painted just before autoclaving, then the coloring cannot be rubbed off and may be touched up from time to time to keep for years. This way the media can be quickly identified whether they are on the shelves or in the water bath, thus helping to eliminate possible costly errors.

It is common practice to set out bottles of assay media the night before the assay so that they can be put in the steamer or autoclave for melting down first thing the following morning. It is advisable to melt down a bottle or two extra in case of any breakages during this process, but these extras must be discarded if they were not needed because repeated heating could damage the nutritive properties of the medium thus affecting the growth rate of the inoculum. Even worse than that, the extra heating could seriously affect the gelling properties of the agar, especially if the medium has a pH of 6.5 or lower. In this case it would be very difficult to remove the agar disks cleanly and without damaging the contours of the wells in the cup–plate assay leading to uneven diffusion.

Liquid media used for turbidimetric assays are also vulnerable to repeated heating, and any residue left after addition to the assay tubes should be discarded. This is especially true in the turbidimetric assay of vitamins and amino acids, because these tend to have complex formulas with many heat-sensitive ingredients.

Slight variations in the composition of media due to bottle differences do not usually lead to problems in the case of large-plate assays, because each plate is poured from only one bottle of agar. However, this could be a source of error in petri dish methods, where several bottles of media may be used for pouring dishes used in a day's assay. Whenever possible, use the same batch of medium on any one day. In any case, record the batch number of the medium used on the worksheet.

The most important aspect of the control of assay media is the validation of each batch made. If a particular medium is in regular demand it is better to prepare a large batch which will last for at least 2–3 weeks. Most media can be stored at room temperature until required, provided the bottles were firmly sealed after sterilization. This is especially true of certain multiuse media which have the same basic composition but differ only in their final pH. These can be readily made up without adjusting the pH, and the final pH can be controlled on the day of the assay just before inoculation. For example, antibiotic medium 1 (see Appendix 4) can be used at pH 6.5, 7.9, 8.3, etc., depending on the actual antibiotic being assayed. Therefore, ensure a continuity of supply by preparing another batch before the current batch has been used up. Before the new medium is used it must be tested to make certain that the freshly prepared medium will perform just as well as the previous supply. Do not test the medium as soon as it is made, but allow the agar to solidify so that it will have to be reheated for the test. In this way the test sample gets the same double heating that the medium would get in the course of a normal assay. Then check the pH and adjust it if necessary as detailed under the appropriate individual assay method.

In the case of diffusion assays the absolute minimum test could be to plate out the appropriate standard solutions in quadruplicate. After incubation check that zone definition is satisfactory and that the value for slope *b* is comparable with values normally obtained. A more convenient check would be simply to replicate one of the complete assays by spotting out all the solutions, i.e., both standard and samples, but using a different random design. This way not only the zone definition and zone sizes are checked but also an extra assay result is generated which can be directly compared to its routine "pair." If the results are within acceptable limits and agree well with the average result of the other two duplicate assays, then the extra-check result can be incorporated in the reported result with added confidence. If the results do not agree with either of the two duplicate results or with their average, the medium check will have to be repeated using a fresh bottle.

It will be necessary to create a logbook in which all the relevant information is entered for each type of medium in use. Suggested headings could be: batch no., date tested, pH found, final pH, type of assay, slope *b* result, and remarks.

Media for tube assays tend to be prepared as they are needed and liquid stocks are not normally kept. This is especially true of the complex media used in the assay of growth-promoting substances. These media are available in dehydrated form. In such cases the most convenient way to ensure that the correct medium was used, is to routinely include either an extra standard or a check sample the potency of which is known. The latter is particularly useful if an extraction is involved.

These considerations should highlight some of the control methods necessary to prevent media failure in routine assays and to ensure uniformity of responses.

7.3 Diluents

Some of the comments made under media apply to a large extent to diluents also. Buffer solutions are often made up in large volumes (e.g., 10 liters or more) and supplied in 15- to 20-liter aspirators fitted with a tap. The containers must be clearly labeled, giving a description of the solution (e.g., pH 7.0, buffer solution, batch number, date prepared, and life of expiry, e.g., "Use before . . ."

In general, if the recipe was followed faithfully, the pH of the buffer solution must be correct. Formulas of buffer solutions are generally much simpler than those of assay media; consequently, they are less likely to be incorrect. In spite of that, planned monitoring of the pH will be necessary.

Do not use paper strips for checking the pH because they are not accurate enough to detect changes of less than 0.3 pH units; rather, use a pH meter which measures pH to two decimal places. The pH meter has to be calibrated and standardized on a regular basis and must be maintained in good working order according to the manufacturer's instructions. A logbook is necessary to record calibration, standardization, and maintenance, giving dates and any other relevant details with initials or signatures.

Buffer solutions should be stored in the refrigerator if they are not used immediately or if large volumes are left over, especially if the solutions are not sterilized. In this case sufficient time has to be allowed to ensure that the buffer solution reaches room temperature before it is used for dilutions. The batch number, the identity, and the pH of the diluent used, if applicable, must be entered on the worksheet accompanying the assay.

The distilled water or deionized water used for preparing the buffer solution quite often contains large numbers of waterborne bacteria which can proliferate rapidly if kept at room temperature, causing contamination on assay plates. Growth can take place even at 4°C, and it is not advisable to store solutions in the refrigerator for longer than 1 week.

The use of sterile buffer solutions is sometimes recommended. If desired, buffer solutions can be sterilized in an autoclave filled in volumes of 2 liters or less or by passing the solution through a sterile 0.22-μm membrane filter into suitable sterile containers.

7.4 Stock Cultures

It was previously mentioned in Chapter 2 (Section 2.4) that each culture must be clearly identified with the species name, ATCC or other type culture collection number, and date of subculture.

When the assay organism is first received from a culture collection, usually in the form of a freeze-dried culture, it is properly labeled. As time goes by there is sometimes a temptation to omit some of the vital details when subculturing, the usual excuse being that the screw top or the test tube is too small to accommodate all the relevant information. The temptation must be resisted. Marking ink is quite satisfactory if a fine-point marker is used. The generic name may be abbreviated. When writing on the glass surface, it is better to write on the part where the agar slant is, so as not to obscure the slant's surface where the growth is. It is important to have a clear view of the growth surface because after incubation the inoculated slant needs careful examination to ensure that the culture shows typical growth without any signs of contamination. An indelible marker should be used because other types of marking can be quite easily rubbed off accidentally during subculturing. Peel-off labels written with a ballpoint pen or typed are neater but again must not obscure the growing surface. Stick-on labels are not recommended as they could create problems during washing.

Assay organisms have to be cultured under optimal conditions and subcultured regularly, otherwise the zone definition deteriorates or the slope of the response in turbidimetric assays changes and will result in inferior assays. Whenever possible, grow up a fresh culture for preparing the inoculum so that the assay organism is kept in its logarithmic phase of growth as far as possible.

Organisms used in the assay of mixtures of antibiotics very often have to be maintained on slants containing the antibiotic to which they are resistant to ensure that this characteristic is not lost. Resistant strains have been known to revert to a nonresistant form when the stock culture has been repeatedly subcultured in the absence of the relevant antibiotic. Equally, some organisms used in vitamin assays also need low concentrations of the appropriate vitamin in their maintenance medium to ensure that their dependence on the vitamin is not lost.

Any noticeable change in the assay performance, whether sudden or gradual—such as zone diameters becoming too small or even nonexistent, the value of slope b becoming too small, or growth in the blank tubes becoming too heavy—can all point to a problem with the assay culture. In these cases the best thing to do is to go back immediately to the submaster culture (see Chapter 2) and examine it thoroughly for signs of contamination. If the latest submaster looks suspicious, discard it and examine the previous one. It is a good idea to keep up to three or four generations in the refrigerator and just discard the oldest one after each satisfactory subculturing. Grow up a fresh slant from the submaster which shows no signs of contamination and test it as explained in Chapter 2 and in Section 7.5. If there is no improvement, it will be necessary to test the master culture or in the last resort to start afresh with a new freeze-dried culture.

Kavanagh and Ragheb (1979) recommend that identity and purity of cultures should be checked at least annually. Examples of possible "mistaken identities" are quoted by Kavanagh (1975a) and Hewitt (1977). New assay cultures should always be checked for purity no matter where they are obtained. Appropriate procedures are described in Chapter 2.

7.5   Inoculum

An important aspect of using assay cultures is to ensure that whenever possible the inoculum is tested before it is put into regular use. Most microorganisms survive as a stock suspension for a minimum of 1 week and some for months at 4°C. Therefore, the safest and the most certain procedure is to prepare a concentrated suspension as detailed under the relevant individual assay method and then standardize it as detailed in Section 2.7. After a little practice and observing standardized procedures it is possible to calibrate a new suspension overnight. Only concentrated suspensions can be stored with any confidence, so that for turbidimetric assays the suspension has to be diluted daily as required. *Staphylococcus aureus* suspension, for example, at a concentration of $10^6 - 10^8$ colony-forming units (CFU) per milliliter will keep for at least 2 weeks, but at a concentration of 100 CFU/ml or less it will die out rapidly after 5 days at 4°C.

Once again proper labeling of the inoculum is essential: e.g., species name, ATCC or other reference number, date of preparation, and clear instructions as to its use, such as "Rate of use 0.5 ml/100 ml," or "Dilute 1 : 100 and use one drop per tube," or simply "Use 1 ml per large plate."

7.6   Spore Suspension

The advantages of using bacterial spore suspensions were fully discussed in Chapter 2. Once optimal conditions are established with a calibrated spore suspension, the suspension can be kept and used for months. Dispense the harvested concentrated spore suspension in small volumes like 10 or 20 ml after pasteurization, and store in the refrigerator. Choose a volume which can be conveniently diluted to, say, 100 or 50 ml, to be used as inoculum. In contrast to nonsporing bacteria, even the dilute spore suspension can be kept in the refrigerator for a considerable length of time. The dilute spore suspension must not be taken out of the refrigerator until the melted assay medium has cooled to the correct inoculating temperature to make sure that the suspension is not left out at room temperature for any longer than is absolutely necessary. Even under these conditions it sometimes happens that the zone definition gradually deteriorates, which is thought to be due to some of the spores germinating and causing nonsynchronized response. Sharpness of zone edges can often be restored if the dilute spore suspension is periodically repasteurized at 80°C for 20 minutes before it is used again. Prevention is always better than having to find a cure, and since pasteurization does not harm the suspension, this can be done monthly, weekly, or even daily, if it is so desired.

Instead of following the full details given in Chapter 2 on the calibration of spore suspensions, the same technique could be adopted as was described in Section 7.2 for the testing of media. Before the calibrated dilute spore suspension currently in use has run out, dilute another stock suspension according to the instructions on the label (as determined when the concentrated suspension was originally calibrated), and pasteurize it. Inoculate a duplicate plate with the test suspension using the same batch of assay medium that was used with the current dilute suspension, and replicate one of the assays. This way the two assay plates can be easily compared after incubation and, if all is well, an extra set of results can be obtained. This "monitoring" of fresh dilutions of the same batch of spore suspension is an important aspect of assay quality control. All experimental details must be fully recorded in line with current good laboratory practices. A looseleaf folder containing the worksheets, experimental details, and all relevant information for each batch will be found useful.

Each bottle containing the spore suspension must be clearly labeled with name and reference number of assay organism, batch number of the spore suspension, date of preparation, recommended "life," and dilution needed for a particular assay. Dilute suspensions should also state the volume to be used per large plate or per 100 ml of assay medium.

7.7 Working Standards

As full a record as possible must be kept of each working standard. For example:

(a) Name and batch or identity number, such as AWS006 (Assay Working Standard)

(b) Date potency was assigned

(c) Source, including manufacturer or supplier, factory lot number, protocol of analysis, purity, impurities, moisture or loss on drying, etc.

(d) Microbiological potency per milligram, including details of assay organism used; mean of how many assays; confidence limits; whether determination of potency was by petri dish, large plate, or turbidimetric assays; how many dose levels were used; details of the reference standard used and what its potency was; whether the potency figure is "as is," calculated to dry or to base, or after drying at given conditions, etc.

(e) Date the AWS was first used

(f) Storage conditions, e.g., "Store at 4°C" or "Store at −20°C"

(g) Expected "life," e.g., "Use before. . . ." (could be based on the known stability data for the substance)

(h) Quantity of AWS in stock, etc.

(i) Date(s) of potency recheck and results, etc.

Some of these points are more fully discussed in Chapter 3.

Storage of AWS is very important and so is their handling. In most cases storage in a small well-stoppered glass vial at 4°C is satisfactory. In some cases frozen storage is necessary, e.g., nystatin. As suggested in Chapter 3, quantities of about 1 g are suitable in most cases because the AWS must be allowed to warm to room temperature before it is weighed. Larger quantities would obviously take much longer to equilibrate to ambient temperature, but in any case, exposing the whole stock of AWS repeatedly over an extended storage period would have a deleterious effect on its keeping quality and consequently on its potency.

Major pharmaceutical companies tend to set up their own stock of working standards. The intended standard material is usually distributed before calibration in approximately 100-mg quantities, which is most convenient for daily use.

If the AWS is not allowed to reach ambient temperature and an attempt is made to weigh a small quantity of it, there could be a danger of a sudden weight pickup due to the so-called dew effect, which means that humidity from the air could condense on the cold powder thereby altering its true potency. Weighing should be done as quickly as possible and no attempt should be made to weigh an exact 50 or 100 mg. There is always a possibility of moisture pickup during weighing, especially if the AWS was dried before weighing or is hygroscopic, or if the laboratory is not humidity controlled. For really accurate assays each sample should be assayed against at least two separately weighed standards or better still, the assays should be repeated on several days using freshly weighed standards each day. This subject will be treated more fully in Chapter 12.

Minimum labeling of each AWS should give name; lot number; potency per milligram, clearly stating whether "as is" or "after drying at 80°C for 3 hours in a vacuum. . . . ", etc.; and date of preparation. Further instructions are to be found in Chapter 3.

7.8   Standard Solutions

Stock solutions of AWS can usually be kept at 4°C for 7 days or in some cases for longer. There are some notable exceptions; e.g., penicillin G stock solution should be kept for not longer than 4 days when performing routine production control work. For more accurate assays it is advisable to prepare a fresh stock solution of penicillin G every day.

Some analysts have been tempted to keep standard stock solutions for their maximum permitted life with the argument that keeping to the same stock solution ensures uniform responses and facilitates comparison of the various samples. The danger with this approach is that if the stock solution was prepared wrongly due to an error in weighing, for example, then all the results generated during the entire period would be wrong. These comments should highlight the need for the greatest vigilance in connection with the standard solutions. It has been stressed several times earlier that results depend on the standard solution being accurately made up. Therefore, it will be seen that preparation of standard solutions cannot be left to a trainee, but needs a responsible and experienced technician or trained analyst.

Furthermore, it is a fact of life that not everyone is a born analyst. Training of assistants is not only helpful but also essential. Unfortunately, routine assays are not precise enough to detect minor discrepancies, which nevertheless may significantly affect assay results. Assays carried out on several days will be necessary. A standard solution prepared by a new person can be treated as a sample and assayed against the standard prepared by an experienced assistant. After a period of training, the average of 10 separate assay results should fall within ± 2% of the AWS potency, or better, with a percentage standard deviation of around 1.0%.

Often it is possible to use physical methods to check the potency of a freshly prepared standard solution, such as an ultraviolet spectrophotometric method or HPLC. These methods should be used whenever possible in addition to microbiological methods. The assistance of analytical chemists should be sought.

More sophisticated three-dose level high-precision assays (described in Chapter 12) can be used, but these methods call for even more expertise and replication on several days will still be necessary.

7.9 Dilutions

It has already been pointed out in Chapter 3 that volumetric pipettes smaller than 5 ml and volumetric flasks smaller than 50 ml should not be used for dilutions. Volumetric pipettes are calibrated to deliver the stated volume of water at 20°C when the pipette is held in a vertical position, drained for a minimum of 3 seconds, and the residual liquid removed by touching the tip of the pipette against the inside wall of the flask. Even under these conditions manufacturers in Britain, for example, guarantee marked volumes only within tolerances of BS 1583, which states that a grade A nominal 5-ml bulb pipette should deliver between 4.985 and 5.015 ml and a grade B between 4.97 and 5.03 ml. Pipettes which have been in use for some considerable time may well be outside the tolerances set by the manufacturer. In one laboratory, among 40 grade B 5-ml volumetric pipettes, 3 were found to be outside the BS limits, when tested according to the method described by Vogel (1962). At the same time two of the pipettes tested were found to be grade A. These findings indicate that considerable dilution errors could be introduced unknowingly if by chance in a series of dilutions two pipettes of opposite extreme limits were used.

One possible way of achieving a constant error would be to use the same 5-ml pipette for the whole set of serial dilutions, rinsing and drying the pipette after each step. Diluting with an error of less than 1 in 500 is difficult even with class A glassware. Lam and Isenhour (1980) constructed a table to show the maximum error caused by specified tolerances in class A glassware, using one, two, or three steps to achieve a millionfold dilution with pipette sizes of 10–25 ml and flasks of 25–100 ml.

Viscosity, density, or surface tension of the sample solution can further affect the precise volume being delivered. This is often unavoidable when, for example, solvents are used for either dissolving or extracting the active ingredient. In these cases it will be necessary to prepare the standard solution in such a way that its final composition agrees with that of the sample. See also Section 3.4.1.

Table 3.2 shows that the smaller the pipette or the flask, the greater will be the error. For this reason it is recommended that pipettes smaller than 5 ml and flasks smaller than 50 ml should be avoided whenever possible. Convenient tables for dilution of standard solutions are given by Hewitt (1977).

Automatic pipettes are becoming more popular because of speed and ease of operation. They also have the advantage that, provided standard and sample solutions are diluted on the same equipment, the errors tend to cancel out. A diluter built on the principle of the Autoturb diluter can dilute any solution with any other without regard to density, viscosity, or vapor pressure of the solutions, and do it with an error of less than 1.0% at constant temperature as described by Kavanagh (1963, 1972, 1982).

What has been said about volumetric pipettes can apply to a large extent to volumetric flasks as well. Modern color-coded volumetric glassware is helpful because it is easier to spot any odd sizes if they get mixed with the wrong volumes.

It is important to keep volumetric glassware in spotlessly clean condition. Pipettes should be stored flat in padded drawers and flasks in cupboards well protected from dust. Flasks are easy to knock over if stored upright on shelves, and they have to be stoppered, plugged, or covered to keep out dust. A much better method is to store them upside down on specially designed shelves with holes to accommodate the necks of the flasks. Storing them this way obviates the need to plug or cover the open end. It is preferable to store the different sizes in separate groups, with the appropriate volume marked for each position.

7.10 Plating-Out Pipettes

As pointed out in Chapter 4, it is very important to ensure that, as near as possible, equal and accurate volumes of solutions are transferred into cylinders or cups in the diffusion assays, and that the plating out of the solutions is completed as rapidly as possible. There are many different makes of pipetters currently available for this purpose. A common feature of these is that the working parts of the pipettes or "samplers" do not come in contact with the solutions and that they are equipped with nonwettable polypropylene tips which deliver a calibrated volume of solution, so long as they are correctly used and regularly serviced. The manufacturers' instructions must be followed.

It is necessary to check the accuracy of these pipetters when they are first put into use and then periodically, such as once a month. Record the weight of a small glass weighing bottle which is fitted with a ground glass stopper, and weigh individual transfers of distilled water into it. Ensure that the stopper is removed for the shortest possible time while the liquid is being dispensed. Record each weight and calculate the weight of each individual transfer. A minimum of 50 weighings is necessary to check the performance of each pipetter initially; this can be reduced to 10 weighings for the monthly monitoring checks. Calculate the mean volume delivered by each instrument together with the standard deviation and the coefficient of variation. If the statistical limits given by the manufacturer are exceeded when the check is carried out by one of the assistants, it is advisable to ask a more experienced person to repeat the weighings. Mean volumes delivered by different instruments can be expected to vary slightly; e.g., one pipetter may dispense a mean volume of 98 μl, another 102 μl instead of the nominal 100 μl. It is often possible to recalibrate the instrument to deliver the accurate volume, but strictly speaking, this should not be necessary unless the error is much greater, because it is more important that the pipetter should dispense a constant volume with high precision than to dispense the nominal volume with a larger error. So long as the same instrument is used for dispensing both the sample and the standard solutions within narrow limits, the instrumental error will cancel out.

This method of weighing each volume delivered by the pipetter can be put to another use, namely for the training of new staff. Trainees will be required to weigh 100 individually dispensed volumes each day until the coefficient of variation drops to below the value claimed by the manufacturer.

It is normal practice in Britain to use a single nonwettable polypropylene tip to dispense all the solutions without rinsing in between into the 64 positions on the 8 × 8 Latin square plates. Good-quality tips used according to the manufacturer's instructions carry over only negligible amounts from one solution to the other. This fact has often been questioned by analysts unfamiliar with this technique.

The following test can be used to check the suitability of new tips or to check that assistants use the pipetters according to instructions. Unknown to the assay staff, the laboratory head should arrange to have some solutions submitted for assay in such a manner that one of the solutions would finish up at double the expected assay concentration after dilution by the assay staff, while another solution, though labeled to contain a similar concentration, would really be a blank buffer solution. These solutions or replicates of these pairs of solutions would then be spotted out against the usual standards following a random design and the plates incubated as normal (as described fully in Chapter 4). If, after incubation, zones are seen in the positions where the blank solutions were plated out, then the tips are either not suitable or they have not been used properly. Of the growth-inhibiting substances, neomycin is known to adhere to surfaces, and if the test just described fails to show zones or traces of activity in the blank positions then the use of a single tip for dispensing all the solutions for an 8 × 8 large plate has been validated. In the case of growth-promoting substances the best one to use would be nicotinic acid. It must be emphasized again that the success of this test depends on the requirement that the assay staff should be unaware of the test being carried out and that the purpose of the test should be revealed to them only on the following day after incubation.

7.11 Miscellaneous Equipment

7.11.1 Water Bath

Water baths are used for a variety of purposes in a microbiological assay laboratory, e.g., (i) cooling hot agar to about 50°C before inoculation for the agar diffusion assay, (ii) heat-treating spore suspensions to kill vegetative forms and possible other vegetative contaminants, and (iii) incubating tubes in the tube assay for growth-promoting and growth-inhibiting substances. In any case the water bath must contain clear water at the right level and at the correct temperature. A water bath fitted with a stirrer is preferable, because it is easier to maintain the correct temperature without significant fluctuations. If the water bath is used merely to cool the hot agar and to hold it at a fixed temperature like 50°C, then a temperature variation of ±0.5°C is acceptable; however, if the water bath is required for manual turbidimetric assays then the variation must not exceed ±0.1°C. The water bath of the Autoturb system has been designed to regulate the temperature to within ±0.01°C. It is safer to check the temperature of the water daily with a mercury–glass thermometer than to rely on the thermostatic control, if fitted.

Even if it is ensured that bottles or tubes do not float in the water, occasionally some medium may spill into the bath or traces of medium may accumulate in the water from the outside walls of the bottles, encouraging growth of thermoduric microorganisms. Do not wait until the water in the bath becomes murky and the inside of the tank slippery. If the bath is in daily use, a monthly cleaning and change of water is recommended. Deionized or distilled water should be used, because tap water will leave insoluble residues on the paddle of the stirrer and on the walls of the tank as the water evaporates.

Keep the water level topped up regularly. The level of water in the water bath should just about cover the top line of the medium in the bottles or tubes without floating any of the containers. Test tubes or small bottles will have to be immersed packed into suitable racks or baskets. The importance of maintaining correct and uniform temperature in assay tubes was fully explained in Chapter 5.

7.11.2 Incubators

Incubators with an even heat distribution are best for incubation of assay plates for the agar diffusion assay. The difference in temperature between top and bottom shelves should not exceed ±0.5°C. The actual temperature reached is not so critical within a degree or so, as long as all the plates have reached the same temperature as quickly as possible. Kavanagh (1975b) gives details of incubators specially designed for assay work.

Fan-assisted incubators are not recommended for diffusion assays, because the edges of the plates tend to dry out too fast, especially if the lids do not fit very well. Even if there are no obvious signs of drying, the reduction of moisture around the edge of the plate can result in larger zone sizes than those in the middle. This effect can be more pronounced with large plates than with petri dishes. On the other hand, small variations of temperature across a large plate are fully compensated by the design. This is not so for petri dishes. The incubator should be fitted with as many shelves as possible so as to allow the incubation of petri dishes in single layers only. When using large plates, leave just enough space between shelves so that they can be placed inside without spilling or knocking. If in an emergency more than one plate has to be placed on the same shelf, do not stack them directly one on top of another, but interspace them with something suitable such as two thin square metal rods, so as to leave an air space between the lid of one plate and the base of the next.

Incubators used for turbidimetric assays can be of the fan-assisted circulation type to ensure that the temperature fluctuation from shelf to shelf does not exceed $\pm 0.1\,°C$ when assaying growth-promoting substances. Growth-inhibiting substances are assays of growth rate which are extremely sensitive to the slightest changes in temperature. For this reason incubators are quite unsuitable, because the tubes have to be warmed up quickly and uniformly, which can be achieved only in a specially designed water bath like the one used in the Autoturb system. In the assay of growth-inhibiting substances the temperature has to be controlled to $\pm 0.01\,°C$. Refer to Chapter 5 for further details.

Keep a mercury–glass thermometer permanently inside the incubator and check daily that the correct temperature is being maintained.

7.11.3 Agar Cutters

The correct design of agar cutters, whether single or multiple, was described in Chapter 4. Thorough cleaning of these cutters is essential, as an invisible thin layer of agar can be left on the cutter's surface after slicing through the medium. After repeated use, the cutter becomes coated with successive layers of agar which dries between successive operations, forming a solid ring. A nylon nailbrush will effectively remove the dried-on layer of agar from the outside of the cutter, especially after soaking in hot soapy water; a small test tube brush will deal with the inside surface. Rinse with purified water followed by a dip in 75% (v/v) aqueous IMS. If only a single manual cutter is used, it is best to wash it in hot water immediately after use and keep it in a small beaker containing 75% (v/v) IMS and a layer of cotton wool at the bottom to protect the cutting edge. If the cutters are not cleaned regularly, they will cut the wells unevenly, resulting in noncircular zones.

The cutters have to be examined periodically for sharpness and uniformity of shape. Heavy-handed use can cause the cutter to hit the glass plate with too much force, which can damage the cutting edge, especially if the cutter was not held in an exactly vertical position in relation to the glass surface. Hand-held cutters can be damaged in other ways as well, for example, by being dropped or by simply rolling off the bench and hitting the floor. The slightest knock can bend the cutting edge slightly out of the true circular shape, so that the well to be cut will also be noncircular, producing misshapen diffusion zones. For these reasons even single cutters should be vertically mounted and not hand-held.

7.11.4 pH Meter

The necessity to control the pH of every bottle of medium used for an assay was already explained in Chapter 4. This can be achieved properly only if the instrument has been carefully calibrated and then kept in a satisfactory working condition. Manufacturers supply calibration and maintenance instructions, which must be followed.

Measuring the pH of hot agar medium requires specialized combined electrodes which are designed to prevent the ingress of agar into the probe. The authors have been using a Pye Ingold pH electrode (catalog No. 405.88 E07, obtainable from Pye Unicam Ltd., Cambridge, England) with satisfaction for the past 20 years. Another suitable electrode that can be recommended is the Philips CE6 electrode, which has to be stored in a 4 M (saturated) potassium chloride solution when not in use.

Before the pH meter is first used, or wherever a new electrode is installed, the instrument must be calibrated using standard buffer solutions. Standard buffer solutions can be made up according to the formulas given in the BP, USP, or other national pharmacopoeias, or they can be bought already prepared from suppliers of laboratory chemicals. The usual procedure is the "two-buffer adjustment," in which the electrode is standardized against two standard buffer solutions. One of the standard buffers selected is invariably the pH 6.88 solution (see below), while the other one is usually a standard buffer close to the range of the intended pH measurement. These calibrations are best carried out at least once a month and have to be duly recorded in the appropriate logbook.

If the pH meter is in daily use, then a daily check should be carried out against a standard buffer solution, the pH of which is nearest to the one being measured. If the pH meter is used only occasionally, then a full two-buffer standardization will have to be carried out each time before the unknown pH is measured and recorded.

The most commonly used buffers are 0.05 M potassium hydrogen phthalate, pH at 20°C is 4.00; and 0.025 M phosphate buffer, pH at 20°C is 6.88. If the pH meter is intended to be used in the alkaline range, an additional buffer solution can also be prepared: 0.05 M sodium borate, pH at 20°C is 9.22.

The electrode must be stored immersed as recommended by the manufacturer either in a suitable buffer solution or in distilled water when not in use to ensure that the electrode does not dry out. An electrode which has been allowed to remain dry for a long period should be conditioned by soaking it overnight in distilled water acidified with dilute hydrochloric acid to about pH 4 or 5 before a calibration is attempted.

Accurate pH measurements are possible in the temperature range of 20°–25°C for most practical purposes. Kavanagh (1972) points out the importance of pH control in the turbidimetric assay of hygromycin B. When it is necessary to measure and adjust the pH of hot media, the temperature-compensating control must be set to coincide with the temperature of the medium being measured.

As soon as the pH measurement is completed, the electrode must be rinsed with warm distilled water before the agar, if present, has had time to set and block the pores. The special Ingold electrode mentioned earlier has a glass sleeve, which has to be kept loose by gently raising and twisting it after each pH measurement of an agar medium to ensure that the electrolytic contact between the electrode and the solution is maintained.

Examine the electrode regularly to ensure that there are no air bubbles trapped inside the glass electrode bulb or in the potassium chloride crystals. Keep the electrode topped up with saturated potassium chloride solution (in the case of the Ingold electrode also containing a drop of silver nitrate solution). The pH meter can be kept in good working condition if the instrument is serviced by qualified service engineers at about 6-month intervals; these engineers should also be called if the instrument develops a fault which cannot be corrected by the laboratory head.

7.11.5   Measuring Equipment

7.11.5.1   Diffusion assays

Hand-held vernier or dial calipers (Fig. 4.10) require very little maintenance apart from the occasional lubrication.

Zones in petri dish assays can be read from the reverse side of the plates using the flat inner part of the jaws. It is best to remove the lid and place the petri dish upside down on a suitable reading box (Fig. 4.11). Hold the caliper with both hands and lay it across the back of the petri plate so that the stationary jaw is lined up with the left-hand edge of the zone. Hold the caliper steady with the left hand to ensure that it does not slip, and move the sliding jaw as necessary with the right hand until it lines up with the right-hand side edge of the zone; then read off the zone diameter to the nearest 0.1 mm. This method of reading is then repeated for each zone and the diameters are recorded, preferably by another assistant. With a bit of practice this reading can be done quite accurately, provided a random design has been used so that the person reading does not know whether the zone being read is a sample or a standard. If upon inspection the zone appears to be slightly elliptical, other readings should be tried through different angles and the best-matching diameters averaged. This method has the advantage that the caliper does not come in contact with the inoculated agar medium.

A more common form of zone-reading calipers in Great Britain, especially for large plates, are those with needle-point extensions. The jaws either have been ground to a fine point or have had long, thin needles soldered on as extensions (Fig. 4.10). A better design is the one recommended by Bolinder (1972), who had thin metal blades soldered to the jaws of the caliper, thereby eliminating errors due to parallax. Needle calipers are generally used to measure zone diameters on the agar surface. The stationary or left-hand side needle is gently touched against the left-hand side zone edge, so that it sinks into the agar very slightly. The sliding point is then brought up against the right-hand zone edge, exactly opposite the stationary needle, and it is allowed to touch the agar at that point. The zone diameter can now be read off the scales to the nearest 0.1 mm and the measurement dictated to a second assistant. This process is repeated for each of the zones in succession in the petri plate or row by row, traveling from left to right across the large plate.

When all the zones have been read for the day, wipe the "needles" clean with a piece of cotton wool or paper tissue saturated with a suitable disinfectant, such as alcoholic Savlon solution, before storing it safely for the night. It is good practice to store these calipers in a lined wooden case specially made for the instrument to protect the "needles" against damage and to protect people such as cleaning staff against injury, which can easily happen if the calipers are carelessly left lying about. Should the needles of the caliper become bent accidentally during the course of reading a set of zone diameters, do not carry on with the reading, but straighten out the needles and then reread the set or large plate. The caliper need not measure to an accurate metric scale, but any arbitrary units can be used as long as the caliper scale is linear and the measurements are accurately reproducible.

In the United States and in Europe the Fisher–Lilly zone readers and projectors are used more frequently. The scale of the zone reader is graduated in 0.2-mm units, but it is quite common to estimate to the nearest 0.1 mm. Scales of projectors are normally in 1.0 mm units, but it is again possible to interpolate intermediate measurements.

It must be emphasized that zone diameters must be measured as accurately as possible, because reading errors usually contribute the largest share of the "biological error" in microbiological assays. Kavanagh (1972) illustrates the point well in Chapter 2 of "Analytical Microbiology," and Kavanagh and Ragheb (1979) state that an error of 0.1 mm in measuring zone diameter in the AOAC chlortetracycline assay caused a 2.5% error in potency. (See also Chapter 4 and Section 7.13.) When zone edges are poorly defined, the measurements will be even less accurate.

The Fisher–Lilly instrument can be equipped with shaft position encoders to give automatic recording of zone diameters (Kavanagh, 1972). However, to eliminate reading bias an instrument like the Autodata digital caliper is necessary. Refer to Chapter 4 for a fuller explanation.

7.11.5.2 Turbidimetric assays

Turbidimetric assays are subject to the same errors as diffusion assays with regard to weighing and dilution of standards and samples. The importance of accurate incubation temperatures has already been emphasized. Kavanagh (1972) and Rippere (1979) pointed out that the quality of the spectrophotometer used for measuring turbidities was more important than was generally recognized. Kavanagh gives a full description of the various automated systems available.

It is difficult to recommend ways of calibrating instruments used for measuring opacities because there are so many different makes. The World Health Organization (WHO) has standards available for calibration purposes, but unfortunately they are not suitable for many instruments. It is quite easy to prepare one's own standards with a little ingenuity, patience, and "trial and error." All one needs is two or three cuvettes the faces of which have been ground with fine-grade glass paper to varying degrees of opacity. Two or three reference points should be adequate. These specially prepared ground cuvettes are kept wrapped in fine paper tissue when not in use. It is best to tabulate readings of these standards, which should be done at least once a month. Slight differences can be expected in the readings, which are affected by the humidity in the atmosphere. The complexities of photometric assaying are well described by F. W. Kavanagh (1972).

7.12 Designs Used in Microbiological Assays

Randomization of replicate positions for standard and sample solutions is important for both diffusion and turbidimetric assays. It is not generally recognized that significant bias may be introduced into the results if (a) randomized designs are not used, (b) the designs used are not truly random, and (c) the designs are either too few or not being used properly. The use of randomized designs is indispensable in the case of Latin square diffusion assays. Their primary purpose is to eliminate the errors inherent in large plates, as explained in Chapter 4. A secondary, but just as important purpose is to eliminate reading bias as far as possible. If the reader knows which zone corresponds to which treatment, this knowledge can influence the reader's judgment subconsciously, resulting in a biased potency estimate. This effect was highlighted in Chapter 4 when discussing the different methods of reading the zone diameters.

The authors had come across assay laboratories where the petri-dish assays were replaced by large plates but the various solutions were simply plated out in column order (i.e., solution 1 in the first column, solution 2 in the second column) not realizing that zone diameters around the edges often tend to be larger than those in the middle of the plate. A variation of this incorrect method has even been published for the antibiotic assay of clinical specimens by Bennett *et al.* (1966).

However, most workers are aware of the fact that statistically correct designs are necessary when large plates are used. Most of them also know that Latin square designs are described in the statistical tables of Fisher and Yates (1963). However, very few people take the trouble to read the instructions for constructing satisfactory designs. Admittedly the instructions given with these tables are more intelligible to qualified statisticians than to others not so specialized. Consequently, some laboratory workers simply copy the designs exactly as given in the tables, not realizing that they were not meant to be used in that form. In fact, the basic designs given there would be too easy to memorize. Some "ready-to-use" assay designs will be found in Appendix 8.

At one assay laboratory the analysts proudly announced that they were indeed using large-plate assays but had only the one design, which they no longer needed to place under the plates because they had learned the positions by heart! Another laboratory had a fair number of designs in stock, but the analyst in charge confessed that one particular design was rather popular with the assistants. The reason for the popularity became obvious upon closer inspection, because the pattern was easy to memorize. In yet another laboratory a large number of assay designs were available, but cross-questioning the assay leader revealed that, instead of following the designs as was explained in Chapter 4, they always plated out solution 1 first in all rows, followed by solution 2 in all rows, etc. These examples are typical of the misuse of good designs.

The question is often asked, how many different designs should be in use in a well-run assay laboratory. Ideally, each design should be used only once and then discarded, because there are so many millions of designs possible in the case of 8×8 Latin squares. From a practical point of view, it very much depends on the number of assays being carried out daily. If only a few assays are required per week, then about 20 different designs should suffice, but if the assay demand is much greater then obviously a greater number will be necessary. In the authors' laboratories 300 different designs are in use.

The designs must be used in a random order. For this reason do not attempt to sort the designs in any particular order. This is best achieved by collecting the designs every day after use and placing them at the bottom of the pile. The designs needed for the day of the assays will be simply taken from the top by each assistant as they are needed. On no account must a particular design be selected out of turn once they have been "shuffled."

The use of randomized designs is equally important for turbidimetric assays. This was not recognized in the early days of vitamin assays, when a separate rack was used for each sample and each standard solution. Five different concentrations in triplicate usually filled each rack. Statistical analysis by Robinson (1971) showed that there was a significant bias between racks which necessitated random distribution of both standard and sample solutions within each rack.

Every rack was given a design number, and every position in the rack was assigned a random number. Suitable designs are given in Appendix 8, and the correct use of these designs is explained in Chapter 11.

The same remarks are also true for the turbidimetric assays of growth-inhibiting substances. Each rack of tubes must be a complete test, the standard and sample solutions arranged according to a random pattern. All solutions, such as medium, treatments, diluents, and inoculum (if not already present in the medium), must be added in the same order, and turbidities also must be measured in the same order. More details will be found in Chapter 11.

7.13   Zone-Reading Performance

It would be a mistake to assume that each and every assistant could read the zone diameters of a diffusion assay with equal precision. It should be well known by now that one of the largest sources of error in diffusion assays is associated with the reading of zone diameters in the case of manual reading methods. For example, if the same large plate is read by several assistants, it will be noted that different people read different diameters for the same zones. This in itself is not significant so long as the replicate observations are consistent. However, there are people who are unable to distinguish between zone diameters of different sizes. In fact, the harder these people try the more mistakes they make.

Statistical analysis of each set of readings gives an important clue. The percentage standard error figure calculated for each large plate, for example, gives an indication of the overall confidence limits. The lower this figure is the better it is. For example, in a well-run assay laboratory one could expect a percentage standard error of around 2.0% on an 8×8 Latin square design, using three samples against one standard and a dose ratio of $2:1$ corresponding to approximate confidence limits of $96-104\%$ ($p = 0.95$). Careful examination of the percentage standard error figure may reveal certain average trends which are associated with individual assistants. Trainees would invariably be expected to produce higher errors, which would be reflected in higher percentage standard error figures. If after training these figures do not decrease to a norm associated with the particular assay laboratory but remain at a high value — often twice the norm, or even higher — then an investigation into the reason for the high values is called for. In most persistent cases the cause will be found to lie in reading error. If the same set of zones is reread by another more experienced assistant whose usual reading performance is equal to or better than the accepted norm, the percentage standard error figure will often show a remarkable improvement. If there is an automated method of zone-measuring equipment available, then this could form the basis of comparison. A few examples of such replicate readings are shown in Table 7.1.

If the second reading does not indicate a significant improvement in the percentage standard value, then the reason for the high error must be sought elsewhere. Other causes could be faulty technique in plating out the solutions or faulty equipment such as a pipetter. The latter could be checked as described in Section 7.10. If after repeated instructions and advice the percentage error figures remain excessive or erratic, and the high value is due to the individual's poor reading performance, then the assistant concerned must be given other responsibilities.

Table 7.1

Comparison of Percentage Standard Errors Calculated on 8 × 8 Antibiotic Diffusion Assays When Zone Diameters on the Same Large Plate Were Read by Two Different Assistants

Example	Assay	First reader	%SE[a]	Second reader[b]	%SE[a]
1	Cefuracetin	KC	3.8	SV	2.4
2	Cefuracetin	KC	5.0	SV	2.9
3	Cefuracetin	KC	4.3	SV	2.6
4	Neomycin	KC	6.3	LMA	4.4
5	Oxytetracycline	NS	2.5	SV	1.5
6	Oxytetracycline	NS	5.0	SV	2.4
7	Oxytetracycline	KC	3.4	LMA	1.7
8	Oxytetracycline	KC	2.5	LMA	1.9
9	Oxytetracycline	NS	15.6	LMA	2.4
10	Oxytetracycline	KC	3.8	SV	2.4

[a] %SE, Percentage standard error.
[b] The second reader is obviously the more experienced one.

7.14 Washing Performance

Washing of reusable glassware is an important activity which must be done properly to ensure that all traces of active substances have been removed. Examples of especially difficult substances to remove are neomycin and nicotinic acid. Old-fashioned methods still insist on using dangerous chemicals like chromic acid for washing glassware. Much safer and much more effective are the present-day detergents like Pyroneg and Decon 90 (See Appendix 1). For best results follow the manufacturer's instructions.

Lightbown (1975) reported the development of "ghost zones" on large plates due to inadequate washing. This effect was shown by pouring a very thin layer of seeded assay agar into the glass plate. After overnight incubation rows of zones developed over the spots where the washing process was unable to remove the last traces of the adhering antibiotic. This is certainly a simple but effective visual method for validating the washing process, but it is important to remember to use a very sensitive microorganism for seeding the assay medium. This residual effect was most noticeable with neomycin and with penicillinase, which also tends to cling to surfaces.

Occasionally it happens in the turbidimetric assays of growth-promoting substances that an inoculated blank (i.e., containing single-strength assay medium only) shows excessive growth, or replicate tubes of the same concentration exhibit uneven growth, or, in the worst case, every one of the tubes including the standards grows to a uniform turbidity and the gradation of the responses disappears. This is often a result of inadequate washing methods. Sometimes the effect is not quite so drastic, but sometimes one tube in the set may give an unexpected turbidity reading which is "out of step," so that it is either too high or too low when compared to replicate tubes.

The following washing procedure can be recommended to overcome these problems, even if an automatic washing machine is available.

(1) Soak test tubes, pipettes, and volumetric flasks in a proprietary detergent solution at a concentration, time, and temperature recommended by the manufacturer (e.g., in Decon 90 at a concentration of 2% overnight at room temperature).

(2) The following day, pour out the dilute detergent solution from all the glassware items and rinse them out *immediately* with copious amounts of tap water.

Note: Detergent must not be allowed to dry on the glassware, but must be rinsed off with plenty of water.

(3) Rinse at least three times with distilled water and allow to drain.

(4) Fill all the test tubes with glass-distilled water, cover with a sheet of aluminum foil (having previously stacked the tubes in a suitable wire basket), and autoclave them at 121°C for 15 minutes.

(5) Allow to cool, then empty and drain the tubes. Transfer the tubes into another wire basket so that they are upside down.

(6) Cover with foil and dry them in a glassware drying cabinet or incubator.

(7) Finally, label the dry tubes "Ready for tube assays."

7.15 Finding the Correct Dose Levels

The authors have often visited laboratories where assays were conducted at concentration levels which were too insensitive to detect small differences in sample potencies. Superficial examination of the raw data indicated that replicate readings were in good agreement and sample potencies also tended to come close to those of the standard. Assay leaders were usually well pleased with the results, not realizing that these might not reflect the true potencies. Most frequently the underlying cause was simply that the inoculum used was too concentrated, which affected the slope of the assay thus making the assay insensitive (see also Section 7.16). This effect was clearly illustrated by Hewitt (1977).

For accurate assays it is important to choose dose levels within the "sensitive region" of the response curve where small changes in concentration give maximum changes in response (e.g., zone diameter, turbidity). The correct inoculum size should be established when the suspension of the assay organism is calibrated as described in Chapter 2. In the case of diffusion assays the secret is to mix in just sufficient inoculum to obtain a uniform, confluent growth after incubation which gives sharply defined zone edges. Addition of more and more inoculum will gradually depress the zone sizes and reduce the slope of the assay, resulting in loss of sensitivity.

Having established the right basic conditions for a high-quality assay, the next requirement is to find the most appropriate dose levels. This is especially important if an assay is required for a brand-new compound. Weigh a suitable amount of the compound and prepare a solution of a suitable concentration. In most cases a 100-μg/ml concentration is high enough to start with. Prepare doubling dilutions in series and plate out each dilution in quadruplicate. It may be necessary to prepare similar doubling dilutions in distilled water and various buffer solutions, depending on the compound under investigation. After incubation, possibly at various temperatures, critically examine the zones produced for sharpness and record the zone sizes for each dilution. It is important to extend the dilution series until zone diameters become too small to measure. Plot the average responses against the logarithms of the concentrations used, and draw a graph. A typical example is shown in Fig. 7.1.

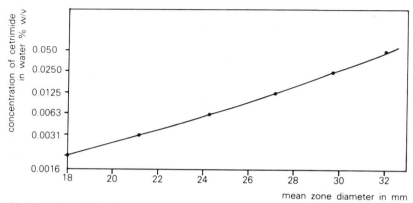

Fig. 7.1. Graph showing average zone diameters of aqueous cetrimide standard over six dose levels plotted against logarithm of concentration using *Micrococcus flavus* (now known as *M. luteus*) as test organism.

It will be noted that for all practical purposes a large part of the graph will give a straight line. Select the middle of the straight-line portion and fix the dose levels around this point. Round off the dose levels to sensible working concentrations (e.g., 1.0 and 0.5 μg/ml, or 0.2 and 0.1 μg/ml). If three dose levels are required, select the three concentrations so that the middle dose lies approximately on the middle of the straight line. This graphic approach will also supply other useful information, such as b, the slope of the straight line, which will determine whether small differences in potency will give significant differences in zone diameter. Whenever possible, assay conditions should be set in such a way that the log dose–response line makes a relatively small angle with the x axis, in which case a $2:1$ dose ratio will be satisfactory. If the angle is too large, then a $4:1$ dose ratio will be required. It should be obvious that a higher dose ratio indicates a less sensitive and consequently less accurate system.

Much of what has been said about the agar diffusion assays will be true of turbidimetric assays as well, but because the actual working range of the latter is generally much shorter, the selection of the working dose levels will be more difficult and much more critical.

7.16 Adjusting Zone Size

The authors came across assay laboratories, even some belonging to well-known pharmaceutical companies, where the zone diameters used were too small to give really accurate results. In a number of cases the dose levels employed were higher than those recommended by the various pharmacopoeial methods. As explained in Section 7.15, they were working in the insensitive region.

The most common reason for this was the use of too heavy an inoculum. Even if the initial inoculum added was correct, leaving the inoculated agar plates at room temperature for an hour or two before applying the test solutions can result in doubling or trebling the original numbers of microorganisms. Figure 7.2 demonstrates this point. It is important to note that not only are the zone diameters decreasing but the slope of the assay is also adversely affected. Sometimes the medium used was too rich, allowing a very fast rate of growth of the assay organism; other times the properties of the agar used were unsuitable (see Section 7.17), in not allowing a fast diffusion of the test substances.

It is therefore highly advisable to exercise extreme care and attention to detail in every assay and in particular to follow the general guidelines and hints given in the various chapters.

If the initial inoculum is suspected to have been too concentrated, simply dilute the initial suspension 1:2, 1:4, 1:5, 1:10, etc., as required, and set up test plates with standard solutions, as described in Chapter 2. After inoculation, agar plates should be poured immediately and should be allowed to stand at room temperature only as long as it is necessary for the agar to set and for the surface moisture to evaporate. Depending on the environmental conditions in the laboratory, this should not take longer than about 30 minutes; the important thing is to adopt a standard time in accordance with the local conditions and keep to a well-established routine.

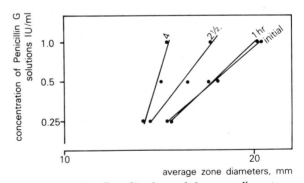

Fig. 7.2. Graph illustrating the effect of leaving seeded assay medium at room temperature before applying test solutions. Not only are zone diameters getting smaller after longer standing, but the slope of the assay is also adversely affected resulting in loss of sensitivity.

Once the right conditions for a good assay have been established, if necessary by trial and error, do not deviate from the steps laid down but keep to the same routine forever more. Detailed standard operating procedures should now be written up to ensure that all assay staff could faithfully reproduce the routine assay. Failure to observe this simple advice will introduce unacceptably large variations from day to day.

7.17 Testing Agar for Suitability in Diffusion Assays

Most assay media can be obtained "ready-made" in dehydrated form, usually only requiring the addition of water to complete the formula. There are a number of reputable manufacturers (see Appendix 2) who sell quality-controlled culture and assay media. The instructions supplied for the preparation of the media must be faithfully observed.

Some assay media described in this book may not be available from commercial sources. Also some assay laboratories may wish to save money by making up media from individual ingredients, especially in countries where local labor is freely available.

One of the important ingredients of media used for diffusion assays is agar. The generally available bacteriological agar is perfectly satisfactory for the preparation of bacteriological culture media but may not be suitable for use in diffusion assays unless certain other criteria are met. The following tests will be required to check the suitability of agar for diffusion assays: pH, gel strength, and diffusion properties. These tests must be carried out on a sample of each fresh batch or purchase of agar side by side with a control sample which has given satisfactory results previously. Over the years the authors tested many samples of agar from different sources, and every batch of Davis Standard Agar (obtainable in 5-kg packs from Davis Gelatine Ltd., Warwick, England) has proved satisfactory in every respect.

7.17.1 pH

Prepare aqueous solutions of agar in three concentrations as described in Section 7.17.2, and determine the pH of each solution before the agar is poured into petri dishes. Use freshly distilled water for the preparation. The pH of the solutions should be pH 5.5 or higher. Should the pH be lower than pH 5.5, it will be necessary to prepare fresh solutions of agar but adjust the water to pH 8.0 before heating.

If the pH of the agar solution is allowed to fall below pH 5.5, the agar layer will become too soft after heating and it will be difficult to remove the agar disks cleanly after cutting with the punch.

7.17.2   Gel Strength

Using freshly distilled water, prepare aqueous solutions of agar from both the test and the control samples as follows. Weigh 1.5, 2.0, and 2.5 g quantities and add 100 ml water to each; soak for 30 minutes, shaking occasionally. Steam the suspensions at 100° – 102°C for 15 minutes, then increase the temperature to 121°C; finally, allow the pressure to drop to atmospheric. Bottle and sterilize the solutions at 121°C for 15 minutes and cool to solidify. Finally, resteam the preparations at 100° – 102°C for 45 minutes and, after cooling to 50°C in a water bath, pour the agar solutions into petri dishes to form a uniform layer of about 4 mm thickness. Let the dishes stand at room temperature for 30 minutes, and then note the clarity of the layers. The agar should be transparent without lumps or undissolved particles. Test the gel strength by gently touching the agar with the fingers; then, using the special agar cutter, cut through the agar layer in several positions and try to remove the resulting agar disks. Score each concentration as "soft" or "firm," and on this basis decide which concentration of agar should be used for the assay media. Retain the "firm" plates for the diffusion test (Section 7.17.3). The heat treatment described above represents the most severe heating process the agar would have to undergo during routine preparation and use.

7.17.3   Diffusion Test

Prepare a 0.1% aqueous saffranin solution and add 0.1 ml of this to each hole cut in the "firm" agar plates (see Section 7.13.2). Allow this to stand on the bench at room temperature for about 24 hours. An unhindered, even diffusion of the red dye around the hole without well-defined zone edges indicates an agar of good diffusion properties. The larger the area penetrated by the dye, the better it is. A small intense red zone with well-defined zone edge and no further diffusion of the dye signifies an agar of poor diffusion properties, which would make it unsuitable for diffusion assay work.

7.18   Staff Selection and Training

It was hinted in the preceding chapters that the so-called biological error often said to be associated with microbiological assay results was due to human carelessness or ignorance, rather than other "biological" factors. An assay result will be only as reliable as the understanding, care, and technical expertise of the people carrying out these assays. Kavanagh (1963) and Simpson (1963) both emphasize the need for staff selection and training of assay staff. Horwitz (1977) discussed some of the important factors necessary for accurate analytical work, and Kavanagh and Ragheb (1979) have also drawn attention briefly to this requirement.

If results produced by microbiological assays are to be meaningful, the work has to be carried out by people who are trained analysts. When interviewing candidates one must look for a keen but steady, relaxed personality endowed with a logical mind, a basic knowledge of bacteriology, an overriding interest in the work they are about to undertake, a willingness to work hard, good powers of concentration, and good practical ability.

Candidates with an absorbing hobby can indicate perseverance and concentration, especially if they managed to pass some recognized examination or achieved some other form of distinction in their subject of interest. Manual dexterity can be often recognized in other nonscientific interests like metal work or interest in electronic kit building. These outside interests can also indicate whether a person is likely to carry on repetitive work successfully without becoming bored with the job.

Having appointed a new member of the assay staff, the long and rigorous training process begins. Points to be stressed to trainees from the outset are that attention to detail at all times is essential, and that the work is likely to be repetitive but not monotonous, because the types of assays to be done often change from day to day, even though a lot of the preliminary preparative work may be identical. Trainees must follow the instructions to the letter, while at the same time maintaining an inquiring attitude. The foregoing sections of this chapter contain many hints and instructions for the sort of training a new microbiological analyst has to receive.

The first day is usually spent in informal discussions with the laboratory manager who outlines the basic principles involved. The trainee is acquainted with the rest of the team, the working conditions, the basic tools and techniques, as well as the safety aspects, good housekeeping, and after-work cleaning operations. The importance of teamwork in certain steps of the assay work is also stressed.

From the second day onward an experienced assay leader or assistant is appointed to guide the new person through the complex and intricate background preparatory operations. Some simple practical tasks are introduced, such as leveling and pouring plates, cutting and "picking" the agar disks, or preparation of test tubes and equipment for tube assays. Replicate assay of a single sample is given on the second or third day in parallel with the appointed tutor. The amount of supplementary information and the number of samples to be assayed are gradually increased over the ensuing weeks. Correct dilution techniques, preparation of standard solutions, reading of results, and calculations are introduced at a later stage. The trainee is encouraged to make copious notes and to ask as many questions as possible. It is useful to draw up a formal training schedule so that individual progress could be monitored. The laboratory manager must find time to talk to the trainee at least once a week so as to keep a watchful eye on the trainee's development, in addition to receiving regular daily reports from the appointed tutor. On these occasions a word of praise and encouragement by the manager for progress made is important. At the same time, any deviation from accepted practice must be corrected at the first opportunity and whenever possible a full explanation given. The full training period depends on the individual but on average can take a minimum of 3 months, or up to 6 months.

An attempt has been made here to show in a nutshell the importance of staff selection and training, and it is hoped that the many practical hints given throughout the book will be of help to laboratory managers in particular. However, it must be pointed out that good analysts are born and not made.

References

Bennett, J. V., Brodie, J. L., Benner, E. J., and Kirby, W. M. M. (1966). *Appl. Microbiol.* **14**, 170.
Bolinder, A. E. (1972). *In* "Analytical Microbiology," Vol. II (F. W. Kavanagh, ed.). Academic Press, New York and London.
Cooper, K. E. (1963 and 1972). *In* "Analytical Microbiology" (F. W. Kavanagh, ed.), Vols. I and II. Academic Press, New York and London.
Fisher, R. A., and Yates, F. (1963). "Statistical Tables for Use in Biological, Agricultural and Medical Research," 6th ed. Longman, London.
Hewitt, W. (1977). "Microbiological Assay." Academic Press, New York.
Horwitz, W. (1977). *In* "Quality Assurance Practices in Health and Environmental Laboratories." American Public Health Association, Washington D.C., Chap. 5.
Kavanagh, F. W., ed. (1963). "Analytical Microbiology," Vol. I. Academic Press, New York and London.
Kavanagh, F. W., ed. (1972). "Analytical Microbiology," Vol. II. Academic Press, New York and London.
Kavanagh, F. W. (1975a). Personal communication.
Kavanagh, F. W. (1975b). *J. Pharm. Sci.* **64**, 1224.
Kavanagh, F. W., and Ragheb, H. S. (1979). *J. Assoc. Off. Anal. Chem.* **62**, 943.
Kavanagh, F. W. (1982). Personal communication.
Lam, R. B., and Isenhour, T. L. (1980). *Anal. Chem.* **52**, 1158.
Lees, K. A., and Tootill, J. P. R. (1955). *Analyst* **80**, 95.
Lightbown, J. W. (1975). Personal communication.
Rippere, R. A. (1979). *J. Assoc. Off. Anal. Chem.* **62**, 951.
Robinson, W. D. (1971). Personal communication.
Simpson, J. S. (1963). *In* "Analytical Microbiology," Vol. I (F. W. Kavanagh, ed.). Academic Press, New York and London.
Vogel, A. I. (1962). "Textbook of Quantitative Inorganic Analysis," 3rd ed. p. 198. Longman, London.

ASSAY DESIGN AND EVALUATION

8.1 Definitions

The dose–response relationship which provides the quantitative basis of any assay, the experimental design of that assay, and its statistical evaluation are, all three, interdependent, and so it is convenient to discuss them together in one chapter.

8.1.1 Dose–Response Relationships

The classical statistical evaluations that will be outlined in this chapter are dependent on the postulation of one of two straight-line relationships: Type 1, most commonly in biological assays in general, the response or some mathematical function of response is directly proportional to the logarithm of the dose. This leads to the "parallel-line assays" that are described in Section 8.2.1. Type 2, less commonly, the response or some mathematical function of the response is directly proportional to the dose itself. This leads to the "slope ratio assays" that are described in Section 8.3.1.

To the first group belong the agar diffusion assays, whether of growth-inhibiting or growth-promoting substances, as well as the turbidimetric assays of some growth-inhibiting substances. To the second group, ideally, belong the tube assays for both amino acids and vitamins. An unusual relationship belonging to this group has been described by Kavanagh (1968): for the turbidimetric assay of some growth-inhibiting substances, the logarithm of the response is the function that is directly proportional to dose itself.

8.1.2 Experimental Design

The parameters that together give a complete description of the experimental design are as follows:

(i) *Number of treatments,* i.e., number of test solutions, which is dependent on (ii) and (iii) below.

(ii) *Number of preparations,* i.e., number of samples plus standard(s).

(iii) *Number of dose levels for each preparation and the relationship between dose levels.* For type 1 dose–response relationships, dose levels should form a geometric progression, but for type 2 relationships they should form an arithmetic progression.

(iv) *Replication of treatments,*   i.e., the number of responses (zones or tubes) to each treatment.

(v) *The manner of physical distribution of the treatments,* e.g., each treatment appearing once in each of six small plates in an agar diffusion assay or once in each row and each column of a 8 × 8 large-plate assay or perhaps in each of three racks in a tube assay.

8.1.3   Statistical Evaluation

The emphasis here is on the word statistical. The classical statistical evaluations for the two types of dose–response relationships are based solely on the internal evidence of the assay. Effectively, through the statistical technique known as analysis of variance (AOV), they determine measures of difference in response due to different preparations, different dose levels, different "blocks" (i.e., different plates, and rows and columns within a plate), or different racks in the case of a tube assay. They also determine measures of deviation from the ideal dose–response line such as curvature and, finally, by difference from the overall variation, the residual variation due to random differences in replicate responses to the same treatment.

A comparison of these measures leads to an assessment of the "validity" of the assay, and from the residual error, fiducial limits or confidence limits for the estimated potency of the sample(s) under test are estimated.

The results of such statistical evaluation can, unfortunately, be very misleading. The tests for "validity" are in fact tests for "invalidity." The fact that in a particular assay these tests do not indicate invalidity is not necessarily indicative that the assay is valid. Moreover, the finding that a particular feature of an assay is statistically invalid does not necessarily mean that there is anything at all wrong with the assay. The meaning of these apparently strange assertions will become clear in Section 8.2.2.

The term confidence limits is itself misleading and may easily result in unwarranted confidence in the potency estimate. Since the statistical evaluation is based only on the internal evidence of the assay, it cannot normally take into consideration errors that might arise in practical operations such as weighings, dilutions, moisture content of standard and sample, sampling errors, and extraction errors. An exception to this general rule is in the case of those elaborate assays in which standard and sample appear in duplicate in a single assay and test solutions are prepared from separate weighings. True confidence in a potency estimate can only be attained through complete control of the quality of the assay as indicated in Chapter 7. Statistical evaluation is just one part of that whole.

8.2 Dose – Response Lines: Type 1

In agar diffusion assays for antibiotics, the log dose – response line is generally taken to be straight. In fact, for the reasons given in Section 4.4 of chapter 4, it is slightly curved. Experience has shown that the straight-line approximation is acceptable for almost all work. Similarly, the agar diffusion assay for vitamins assumes a straight-line log dose – response relationship. The full log dose – response line of tube assays for growth-inhibiting substances is sigmoid, but sometimes the central part of the line is near enough to straight to be used directly. In other cases, however, it is necessary to modify responses using some mathematical device such as the probit transformation. These dose – response lines and transformations were described in Sections 5.3 and 5.3.1 of Chapter 5. Since the type 1 dose – response line is much more commonly encountered in the case of the agar diffusion assay than in tube assays, the discussion that follows is directed mainly to the former group.

8.2.1 Designs for Type 1 Assays

In a properly designed assay of this type, the different dose levels for each preparation (standard and samples) are always arranged in a geometric progression when there are more than two dose levels. This results in logarithms of doses forming an arithmetic progression, which greatly facilitates the calculation of potency estimates and statistical evaluation of results. Thus, doses in the geometric progression 1, 2, 4, 8 would correspond to logarithms of doses forming the arithmetic progression 0.000, 0.301, 0.602, and 0.903.

When a sample is to be assayed it is first allotted a nominal or assumed potency. Test solutions are prepared on the basis of the nominal potency. Upon completion of the practical aspects of the assay, observed responses may be plotted against logarithm of nominal dose for the sample and against logarithm of actual dose for the reference standard. It is an initial assumption that these two response lines should be straight and parallel. If sample potency corresponds exactly with its nominal potency, then the two lines should coincide. When sample and standard are of differing potencies, then that difference is indicated by the distance between the two lines along the logarithm of the dose axis. Such assays are known as parallel-line assays.

Some other terms that will be used in the description of this type of experimental design will now be introduced.

Symmetry. A parallel-line assay is said to be symmetric when the number of dose levels, the ratio between adjacent dose levels, and the number of responses to corresponding treatments (test solution) for each preparation (standard and samples) are identical. All other parallel-line assays are *asymmetric.* There is no theoretical limit to the number of samples that may be compared with a standard in an individual assay. In practice, however, the number is limited by the size of the assay plate and the number of zones it can accommodate. Symmetric assay designs are inherently the more efficient in that for a given total number of zones the symmetric assay will always give narrower confidence limits than an asymmetric assay, carried out under otherwise identical conditions.

Replication. This generally refers to the number of responses to an individual test solution in a single assay; however, the term may also be used to indicate the number of times that an individual sample is assayed. The sense in which the word is used should be apparent from the context.

2 + 2, 3 + 3 designs. The 2 + 2 assay design has two dose levels of standard and two dose levels of each sample. The ratio of high dose to low dose is the same for each sample as it is for standard. The assay may include only a single sample or two or more samples compared with the standard. Similarly, the 3 + 3 assay has three dose levels of standard and three dose levels of each sample. The ratios of high to medium doses and medium to low doses are the same as one another and are the same for sample as for standard. Similarly, designs such as 4 + 4 and 5 + 5 have been described. Although such designs are not generally appropriate for the agar diffusion assay, they may be used in turbidimetric assays; the reasons for this distinction will become apparent when dose ratios are considered. A 5 + 1 assay design is one of those described in the U.S. Code of Federal Regulations (CFR). It employs five dose levels of standard and one dose level of each sample. It differs from all other assays described in this book in that each assay plate (petri dish) does not include all test solutions. Instead, each plate includes three zones corresponding to the middle dose of the standards, henceforth known as the "reference response," and three zones corresponding to one other test solution which may be of one other dose level of standard or the single dose level of a sample. Observed mean responses for each preparation on each plate are then calculated. The mean responses to the nonreference test solutions are then adjusted and expressed relative to the grand mean of all reference responses. The purpose of this complex procedure is to correct for differences between individual plates. Such a correction procedure is not necessary when each plate contains zones for all test solutions so that each plate is a self-contained assay. It should be noted that the CFR also describes the 2 + 2 and 3 + 3 assays. The 5 + 1 assay design is not favored in Europe. The 5 + 1 assay design is also described for turbidimetric assays by the CFR.

Dose ratio. This term, applied in parallel-line assays, refers to the ratio between adjacent dose levels. The dose ratio is commonly 2 : 1 when two or three dose levels are employed but may also be 4 : 1 when only two dose levels are employed. A 4:1 ratio would generally be too wide if three dose levels were employed, as the overall range would then be 16:1 and curvature of the response line would become not only more apparent but a greater problem. However, if the slope b (which will be defined later) is very low, then a 10:1 dose ratio may be used in a two-dose level assay.

Latin square design. This is a pattern which has general applications (particularly in agricultural research) and is used in microbiological assay to define the positioning of individual test solutions on large assay plates, e.g., square plates of 30-cm sides that can accommodate up to 64 zones arranged in eight rows and eight columns. In general, a Latin square design is an arrangement of a series of consecutive numbers from 1 to n in the form of a square consisting of n rows and n columns. Each of the n numbers appears once, and once only, in each row and each column. The replication of each number in a single square is n.

Latin square designs may be 4 by 4 (generally written as 4×4), 5×5, 6×6, etc. Commonly used for large-plate assays are the 6×6 and the 8×8. The International Pharmacopoeia (second edition) gives an example of a 6×6 Latin square design for a three dose level assay $(3 + 3)$ accommodating one sample. The European Pharmacopoeia, (Vol. 2) gives an example of a 9×9 Latin square design for a three-dose level assay $(3 + 3)$ accommodating two samples.

The 6×6 Latin square design may be used to accommodate two samples and one standard in a two-dose level assay $(2 + 2)$. The 8×8 Latin square design may be used to accommodate three samples and one standard in a two-dose level assay $(2 + 2)$. It may also be used for one sample only; here there are three possibilities:

(i)   Two dose levels may be used and the replication doubled by giving each test solution two numbers. Pairs of numbers such as 1 and 5, 2 and 6, 3 and 7, and 4 and 8 would ensure that each test solution appears, appropriately spaced, twice in each row and each column.

(ii)   Four dose levels may be used; however, a 2:1 dose ratio would correspond to an overall dose range of 8:1, at which range bias due to curvature of the response line would become appreciable unless sample potency was close to standard potency. The evidence for this is shown in Section 8.2 of this chapter. Such an overall dose range is nevertheless specified in the preferred method of the British Animal Feeding Stuffs Regulations (Anon. 1985). This was the method adopted by the European Economic Community and published in the Commission Directive 72/199 (Anon. 1972). This published method does not *require* that large plates and Latin square design be used.)

Returning to the question of accommodating four dose levels on an 8×8

Latin square design, an alternative is to use a 3 : 2 dose ratio, thus leading to an overall dose range of $1.5^3 : 1$, i.e., $3.375 : 1$. This leads to the practical difficulties of making the dilutions, which cannot be done with standard volumetric flasks and pipettes.

(iii)   Three dose levels may be used by means of a design introduced by Vincent (1983). In this design, known as the HMML design, the mid-dose test solutions are allocated two numbers and so have a replication of 16 on the plate, whereas the high- and low-dose test solutions have a replication of 8. In this way the full capacity of the plate to accommodate 64 zones is utilized.

The 12×12 Latin square has been used for work of the highest precision such as the establishment of working standards by calibration against a master standard. Two weighings and sets of dilutions are made from each of the two preparations (sample and standard). From each of the four primary solutions, test solutions at three dose levels are prepared. Thus, there are 12 treatments which are then allocated numbers from 1 to 12, causing each treatment to appear once in each row and each column.

Quasi-Latin square design. This is somewhat similar to the Latin square but differs in that the series of consecutive numbers is from 1 to $2n$; these are distributed between n rows and n columns. Each number appears once, and once only, in one-half the rows and one-half the columns. In such a design, n must be an even number. The replication of each number in a single square is $n/2$.

Randomization. Some degree of randomization in the treatment of test solutions is generally necessary to minimize the influence of extraneous factors that could lead to bias in estimation of potency. Responses (zone sizes) are dependent not only on the applied dose of active substance but also on physical factors such as time of application to the plate, position on the plate, and temperature. These factors are often interrelated. Thus, position of a test solution on an assay plate may influence diffusion time prior to incubation, since all test solutions cannot normally be applied at the same instant. Position on the plate could again be an influence on response during incubation as a result of uneven warm-up of the plate leading to effectively different prediffusion times and incubation times. Differences in thickness of agar between plates and in different parts of a single plate would lead to differences in zone size. From these and other considerations it follows that if assays are carried out in a regular manner such that, for example, the standard solutions are always processed (applied to the plate) first and sample number 3 test solutions are always processed last, then observations and the resulting potency estimate will be subject to bias. Naturally, every effort must be made to minimize such bias by use of good facilities and good techniques. However, some unwanted influences will inevitably remain. These may be balanced out to a large extent by randomization. Thus, for example test solutions should be applied to petri dishes in accordance with a different design for every plate in an individual assay. The numbers in a Latin square design may be randomized in each row and each column.

8.2.2 Curvature of Type 1 Dose-Response Lines

The effect of curvature of the dose-response line in the case of the agar diffusion assay for antibiotics was reported by Hewitt (1977, 1981). The practical aspects of the study were based solely on the antibiotic agar diffusion assay, because curvature seemed to be a matter of general concern to analysts and because curvature (though generally slight) is a routine fact when the observed response (the zone diameter) is used directly in the calculation of the potency estimate. However, theoretical aspects of the study indicate that findings would be qualitatively applicable to all type 1 assays. Referring again to the antibiotic agar diffusion assay, it was mentioned in Section 4.4 of Chapter 4 that the square of the zone *width* was directly proportional to the logarithm of the dose, and that it followed that the zone *diameter* (unsquared) could not also be directly proportional to the logarithm of the dose. Thus the fact of curvature of the line zone diameter versus log dose is established.

It was shown by some simple algebra that when using the standard procedures for calculation of potency estimate for a *symmetric* assay, a component of quadratic curvature did not bias the calculation of potency estimate at all. In other words, lines of the two forms

$$y = a + bx$$
$$y = a + bx + cx^2$$

behaved identically.

However, the same did not apply for quadratic curvature and *asymmetric* assays, nor did it apply to higher degrees of curvature whether or not assays were symmetric. Practical evidence was found in the case of agar diffusion assay for four antibiotics (streptomycin, tetracycline, phenoxymethylpenicillin, and ampicillin) that the major component of curvature was quadratic. The interaction between the higher degrees of curvature and different assay designs was studied. The result of the interaction was qualitatively the same in all four cases and is shown for streptomycin in Fig. 8.1, in which bias (percentage by which the estimated potency deviates from the known true potency) is plotted against true potency for several assay designs.

It is evident from the graph that for symmetric assays bias is least with a low dose ratio (2 : 1) at two dose levels only, is greatest with a high dose ratio (4 : 1) at two dose levels only, and is intermediate when three dose levels

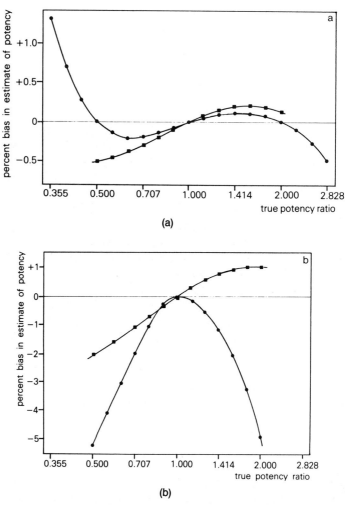

Fig. 8.1. Graphs illustrative of the extent of bias in the potency estimate due to curvature of the log dose–response line. These graphs are based on observations in the assay of streptomycin using *Bacillus subtilis*. Bias is dependent on the parameters of experimental design as well as the true ratio of potency of preparation of unknown potency to that of reference standard. (a) ●———●, 2 + 2 Assay, 2:1 dose ratio; ■———■, 3 + 3 assay, 2:1 dose ratio. (b) ■———■, 2 + 2 assay, 4:1 dose ratio; ●———●, 5 + 1 assay, 5:4 dose ratio.

span a range of 4:1. Bias is greatest for the asymmetric assay illustrated, the 5 + 1 assay with a 5:4 dose ratio, and follows a characteristic pattern as true dose ratio (sample/standard) varies. This pattern, in which bias was always negative, was the same for all four antibiotics. In all cases, bias was at a minimum when true potency ratio approached 1.00.

It is also evident that when a symmetric assay design is used and potency ratio (sample/standard) is not very far removed from 1.00, then bias is of little practical importance. Taking for example a true sample/standard potency ratio ranging from 0.5 to 2.0, then the maximum biases calculated were as shown in Table 8.1 and were small compared with the statistically calculated confidence limits of about $\pm 4\%$ ($P = 0.95$). Although this statistical evaluation had also shown that curvature was "highly significant" (see Section 8.2.3), it is clear from the low bias resulting from the curvature that it is of little practical importance.

Table 8.1

Maximum Biases in Potency Estimates When True
Sample Potencies Lie in the Range 0.5–2.0
Relative to the Standard[a]

	True potency ratio			
Dose ratio	0.500	0.630	1.414	1.782
2 : 1		−0.22%	+0.11%	
4 : 1	−1.85%			+0.95%

[a] These biases, expressed as percentage deviation from the true potency, are as calculated for two-dose level symmetric assays having dose ratios of 2 : 1 and 4 : 1, respectively. These data are based on the work of Hewitt (1977, 1981).

It was stated in Section 8.2.1 that an initial assumption is that response lines should be straight and parallel. Indeed, the calculation procedures demonstrated in Sections 4.6.6 and 5.5.4 were developed from this assumption. It has now been shown that curvature does not necessarily invalidate the assay. It is seen readily from Fig. 8.2 how curvature may lead to apparently nonparallel responses.

8.2.3   Design and Evaluation of Type 1 Assays

In order to appreciate the merits of the various experimental designs that are available for microbiological assay, it is necessary to have some understanding of the principles of assay evaluation. Cognizant of the fact that many scientists and technicians who are biologically orientated are not always mathematically orientated, an endeavor is made here to present assay evaluation in basic mathematical terms. For a more thorough treatment of this topic, the mathematically inclined reader may consult the works of Burn (1950), Emmens (1948), and Finney (1964). The reader who would like more varied examples explained in fairly basic terms is referred to the publication of Hewitt (1977). The method which will be

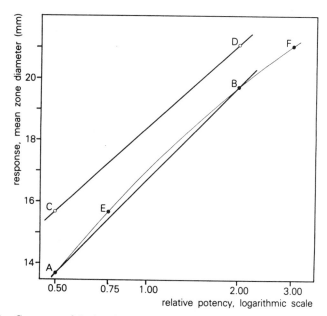

Fig. 8.2. Curvature of the log dose–response line may lead to apparent deviation from parallelism, as illustrated in this theoretical example in which the preparation of unknown potency is "estimated" to have a potency ratio of 1.5 relative to the reference standard. Points A and B are plots of the coordinates of actual log dose and response to the reference standard. Points C and D are plots of the coordinates of nominal log dose and "actual" response to the preparation of unknown potency. After the potency of the unknown preparation has been estimated, the coordinates of estimated log dose and response are plotted as points E and F leading to the curved log dose–response line AEBF. Although the lines A–B and C–D are not parallel, the assay may be perfectly valid. ●———●, True or estimated log dose versus response and O———O, nominal log dose versus response.

illustrated here is not identical with those appearing in well-known pharmacopoeias, although it yields identical results. It has the advantage that it does not use mathematical expressions that are "tailor-made" for a particular assay design (and consequently, not readily adaptable to other designs). Once the principles of the calculation shown here have been grasped, they may be adapted readily for any assay design. Reference has been made in Section 8.1.3 to analysis of variance (AOV). This procedure is illustrated in Example 8.1 of section 8.2.4 for the case of a three-dose level assay using a 6×6 Latin square. Variation in response that can be attributed to known factors are segregated. These are variation due to preparations (standard and sample), dose level (slope of the combined standard plus sample response lines), parallelism (deviations from parallelism between standard and sample response lines, or differences in slope), curvature (i.e., the combined curvature of standard plus sample response lines), opposed curvature (i.e., difference in curvature of the standard and sample response lines), rows (each row contains each of the six treatments or test solutions), and columns (each column contains each of the six treatments).

The calculation of the values that are attributed to these known factors is done by obtaining a figure for the sum of squares of deviations from the overall mean, generally referred to simply as "squares." The value of the "squares" is then divided by the corresponding number of "degrees of freedom" to give the "mean squares." A further component of variation is then obtained by calculating the overall variation "squares" and subtracting from it the sum of the "squares" attributable to each known source. The difference is known as the "residual error squares." This figure is then divided by its corresponding number of degrees of freedom (d.f.) to obtain the "error mean squares." The figure for error mean squares is given the symbol s^2. It is used in the calculation of the confidence limits of the assay. The procedures for calculation of squares and allocation of degrees of freedom are shown in Example 8.1.

To demonstrate the procedure for the estimation of fiducial limits it is necessary to introduce three simple mathematical expressions. These, which are given without showing their derivation, are applicable to all parallel-line assays.

$$V(M) = \frac{s^2}{b^2}\left[\frac{1}{N_S} + \frac{1}{N_T} + \frac{M^2}{S_{xx}}\right] \tag{8.1}$$

in which M is the logarithm of the ratio of sample potency to standard potency, $V(M)$ is the variance of M, s^2 is the residual error, b is the slope of the log dose–response line expressed as the mean increase in zone size corresponding to a 10-fold increase in dose, N_S is the number of responses to all standard doses, N_T is the number of responses to all sample doses, and S_{xx} is a term related to the dose ratio, the number of dose levels, and the replication.

In fact, when M is small (as it is in well-planned assays because sample potency is not far removed from standard potency), then the value of the term M^2/S_{xx} is quite negligible and can be ignored for the present purpose, which is to show the relationship between assay design and evaluation. However, for completeness, the calculation of S_{xx} is shown in Example 8.1.

Having obtained the variance of M, the standard error of M denoted by the symbol s_M, is calculated quite simply as:

$$s_M = [V(M)]^{1/2} \tag{8.2}$$

From the standard error of M, the logarithms of the percentage fiducial limits are then obtained by

$$\log \text{percentage fiducial limits} = 2 \pm t s_M \qquad (8.3)$$

in which 2 is log 100, and t is Student's t, a number that is obtained from statistical tables according to the number of degrees of freedom for residual error and the required probability of the estimate. (t is generally quite close to 2 for a probability of $p = 0.95$.)

Finally, the percentage fiducial limits are obtained by taking antilogarithms.

Having presented the three mathematical expressions which lead to the estimation of fiducial limits, it is now a simple matter to understand how the various features of assay design and, to a lesser extent, practical techniques will have a bearing on the width of the confidence limits. Thus, it is possible to define desiderata in assay design.

Starting from Eq. (8.3), then working backward, it is clear that for higher precision (narrower fiducial limits), both t and s_M should be small. For a probability of $p = 0.95$, the value of t is 2.23 for 10 d.f. dropping to 2.02 for 40 d.f., and to its limiting value of 1.96 when degrees of freedom are infinity. However, at lower numbers of degrees of freedom, the value of t increases rapidly to a maximum of 12.71 when d.f. is 1. The number of degrees of freedom is only low if there is a very low degree of replication; it is rarely less than 15 in typical assays. The way in which degrees of freedom is found will be seen in Examples 8.1 and 8.2.

Turning now to Eq. (8.2), it is seen from the square root relationship that, for example, to reduce the value of s_M by a factor of 2, it would be necessary to reduce the value of $V(M)$ by a factor of 4. This might be achieved by quadrupling the replication, which is the basis of the crude "rule of thumb" that precision (width of fiducial limits) is inversely proportional to the square of the degree of replication—the degree of replication being reflected in Eq. (8.1).

Considering now Eq. (8.1), it is clearly desirable that:

s^2 should be small,
b should be large,
N_S and N_T should be large,
M should be small, and
S_{xx} should be large.

Taking these in turn and considering how these desiderata may be attained or at least approached, the following conclusions are reached:

s^2 is dependent largely on assay conditions and techniques but also, to some extent, on assay design. Considering first the agar diffusion assay, if practical techniques are such that the zone edges are sharply defined and zone sizes are measured by conscientious and well-trained operators having good measuring equipment and good eyesight, then s^2 will be minimized. Now, considering tube assays (which are perhaps invariably turbidimetric in the case of growth-inhibiting substance assays, which are parallel-line assays) then s^2 will be minimized by application of the techniques already recommended in Chapter 5. The value of s^2 is also dependent on degrees of freedom; for a doubling of replication a doubling of the value for residual error "squares" may be expected, however, the degrees of freedom for residual error would be more than doubled so that the net effect would be to reduce somewhat the residual error "mean squares."

b, the slope, is solely a function of the practical assay system and is not in any way dependent on assay design. Its numerical value is unrelated to the units in which dose is measured but is dependent on the measure of response. Thus, if the observed response is zone diameter measured in millimeters and this is used directly in the calculation of potency estimate without modification, then b might, typically, have a value of about $8-12$ (mm). The way in which slope may be influenced practically is discussed in Chapter 4 for the agar diffusion assay and in Chapter 5 for the tube assay of growth-inhibiting substances.

N_S and N_T refer, respectively, to the total number of responses to the standard preparation (whether there is one or more than one weighing) and similarly to the total number of responses to *any one* sample in an assay. It is readily demonstrable that for any individual value of the sum $(N_S + N_T)$, the value of $(1/N_S + 1/N_T)$ is at a minimum when $N_S = N_T$. It follows that symmetric assays are intrinsically more efficient (capable of greater precision) than are asymmetric assays.

M is dependent on good assay planning. If the analyst already has a good idea of the expected potency before setting up the assay, then weighings and dilutions can be arranged so that potency of sample test solution is close to that of standard test solution. This minimizes the value of the term M^2/S_{xx} and thus also the value of $V(M)$ as found by Eq. (8.1). It was shown in Section 8.2.2 that a further reason for keeping the value of M low is that bias due to curvature of the log dose–response line is also minimized.

S_{xx} is the symbol used to denote "the sum of the squares of the deviations of each individual log dose from the overall mean of all log doses." It follows that S_{xx} increases with wider overall dose range and with replication. However, as shown in Section 8.2.2, an overall wider dose range increases the bias due to curvature of the log dose–response line.

To summarize, an appreciation of the three Eqs. (8.1), (8.2), and (8.3) helps in selecting experimental designs that are appropriate to the analysis to be done and also helps in understanding how to endeavor to control the practical aspects of the assay to achieve most efficient results.

The ideals that are concluded by consideration of individual terms in these equations are not necessarily mutually consistent. For example, the slope b may be increased in an agar diffusion assay by reducing the inoculum level, but if this level is reduced too much then zone boundaries will become less sharply defined and more difficult to read; and so s^2 will increase. Also M^2/S_{xx} can be made smaller by increasing the overall dose range; however, this will increase bias arising from curvature. Clearly, compromises based on practical evidence must be made. Apart from the calculation of fiducial limits, the statistical evaluation is intended to show the significance of certain aspects of the data derived in the assay. This is done by a comparison of the "mean squares" or "variances" attributable to individual factors with that attributed to random error. Variance ratios are calculated by dividing mean squares for a particular feature of the assay with the mean squares for residual error. Expressed in its simplest form, the higher the variance ratio, the more likely is the responsible feature to be significant. However, significance in the statistical sense is assessed by reference to tables which take into consideration the number of degrees of freedom associated with each source. This will be illustrated in Examples 8.1 and 8.2.

In a typical assay the variance ratio tests would be expected to show significance attributable to various sources as tabulated below. The list of sources of variation is not exhaustive, but covers all those sources that are encountered in the most commonly used designs.

Source of variation	Significance of the source
Dose level	Highly significant[a]
Difference in slope of response lines (parallelism)	Not significant or of low significance
Curvature of response lines	Not significant or of low significance
Difference in curvature of response lines between preparations	Not significant or of low significance
Difference between preparations	Preferably of low significance but not essential.
Difference between plates in a small-plate assay	Not significant would indicate good standardization of technique
Difference due to rows or to columns in a large-plate Latin square assay	Low significance

[a] In most microbiological assays this is so highly significant that the test is not normally applied; a very low slope such as would cause a less than "highly significant" score would be immediately obvious from the absolute value of b. It is this actual value that is of real importance and not its "statistical significance" as compared with the random error of the assay.

8.2.4 Evaluation of a 3 + 3 Assay

The evaluation of microbiological assays is very adequately described elsewhere and so it is not the authors' intention to duplicate these publications. However, the present work would not be complete without one example of the calculations involved in the evaluation of assays. The example given here is from the work of Vincent (1986) and is an extension of Example 4.1 of Chapter 4, the assay of zinc bacitracin.

Example 8.1: Evaluation of a Three-Dose Level Assay in Which Test Solutions Were Applied to the Plate in Accordance with a 6 × 6 Latin Square Design

This example refers to the assay of bacitracin described in Example 4.1 of Chapter 4. To reduce the labor in computation, observed zone diameters were modified by multiplying by 10 then subtracting 150. These modified responses are tabulated first in the order that they were originally recorded (i.e., in columns corresponding to the six test solutions) and then in a second tabulation corresponding to their actual position on the plate according to the Latin square design. The Latin square design is shown in Fig. 4.4.

Modified responses tabulated according to test solutions:

		Dose level			
	Sample			Standard	
High	Medium	Low	High	Medium	Low
69	50	16	69	49	18
67	47	17	71	44	17
68	46	21	70	44	20
61	41	17	62	40	15
66	41	15	60	40	11
53	30	6	54	33	6
384	255	92	386	250	87

Modified responses tabulated according to their positions on the assay plate:

						Row totals
49	69	18	16	69	50	271
47	17	44	67	17	71	263
70	46	68	20	44	21	269
17	61	41	40	62	15	236
66	15	60	41	11	40	233
6	33	6	54	30	53	182
255	241	237	238	233	250	1454

From the tabulated data, various parameters are calculated as follows:

(i) *Total deviation squares:* S_{yy}

$$S_{yy} = (69^2 + 50^2 + 16^2 + 69^2 + \ldots + 33^2 + 6^2) - \frac{1454^2}{36}$$

$$= 74,402 - 58,725.44 = 15,676.56$$

(ii) *Treatment squares:*

$$= \frac{384^2 + 255^2 + 92^2 + 386^2 + 250^2 + 87^2}{6} - \frac{1454^2}{36}$$

$$= 73,335 - 58,725.44 = 14,609.56$$

The treatment squares figure, which has five degrees of freedom, is then broken down into its five component parts, each with one degree of freedom, by means of orthogonal polynomial coefficients as shown in the tabulation below.

Steps	Polynomial coefficients						$e_i{}^a$	$T_i{}^b$	$T_i^2/6e_i$
Preparations	+1	+1	+1	−1	−1	−1	6	8	1.78
Regression	+1	0	−1	+1	0	−1	4	591	14,553.38
Parallelism	+1	0	−1	−1	0	+1	4	−7	2.04
Curvature	+1	−2	+1	+1	−2	+1	12	−61	51.68
Opposed curvature	+1	−2	+1	−1	+2	−1	12	−7	0.68
Treatment totals	384	255	92	386	250	87			14,609.56

[a] e_i is the sum of the squares of the polynomial coefficients in the corresponding row.

[b] T_i is the sum of the products of the individual polynomial coefficient in the row with their corresponding treatment total.

(iii) *Row squares:*

$$= \frac{271^2 + 263^2 + 269^2 + 236^2 + 233^2 + 182^2}{6} - \frac{1454^2}{36}$$

$$= 59,680 - 58,725.44 = 954.56$$

(iv) *Column squares:*

$$= \frac{255^2 + 241^2 + 237^2 + 238^2 + 233^2 + 250^2}{6} - \frac{1454^2}{36}$$

$$= 58,784.67 - 58,725.44 = 59.23$$

The AOV is summarized in the next tabulation, from which it may be seen that the sum of the squares of the individual components of treatment squares is the same as the figure calculated separately.

Summary of analysis of variance:

Source of variation	Degrees of freedom	Squares	Mean squares[b]	Variance ratio
Preparations	1	1.78	1.78	0.67
Regression	1	14,553.38		
Parallelism (deviations)	1	2.04	2.04	0.77
Curvature	1	51.68	51.68	19.42
Opposed curvature	1	0.68	0.68	0.26
Sub total	5	14,609.56		
Treatments	5	14,609.56		
Rows	1	954.56		
Columns	1	59.22		
Residual error[a]	20	53.22	2.66	1.00[c]
Total (all sources)	35	15,676.56		

[a] Residual error by difference
[b] Mean squares = squares/degrees of freedom
[c] 1.00 by definition as all other mean squares may be compared with error mean squares to assess their significance.

To proceed to the calculation of fiducial limits, the parameters that are to be substituted in Eq. (8.1) are summarized:

$s^2 = 2.66$ (from the AOV)

$b = 81.80$ (from Example 4.1, but multiplied by 10 to conform with the modification of observed responses used in this calculation)

$M = 0.003264$ (from Example 4.1, no modification)

$S_{xx} = 2.175$. This is the sum of the squares of the deviations of individual log doses from the mean log dose and, in this case, is calculated as

$$S_{xx} = 12[(+0.301)^2 + (-0.301)^2] = 2.175$$

Substituting these values in Eq. (8.1) we get

$$V(M) = \frac{2.66}{81.8^2}\left[\frac{1}{18} + \frac{1}{18} + \frac{(0.003264)^2}{2.175}\right]$$

$$= 0.0003975[0.111111 + 0.000005] = 0.00004417$$

From Eq. (8.2),

$$s_M = (0.00004417)^{0.5} = 0.006646$$

The logarithm of percentage fiducial limits is obtained from Eq. (8.3) by substituting $s_M' = 0.006646$ and $t = 2.09$ (obtained from tables for $p = 0.95$ and 20 d.f.), thus log percentage fiducial limits are

$$2 \pm 2.09 \times 0.006646 = 1.9861 \text{ to } 2.0139$$

The corresponding percentage fiducial limits are 96.9 to 103.3% ($p = 0.95$).

In Example 4.1 the estimated potency was calculated to be 55.4 IU/mg; applying these fiducial limits the potency may now be reported as 55.4 IU/mg, limits 53.7–57.2 IU/mg ($p = 0.95$). Consideration may now be given to the meaning of the tabulated "variance ratios," i.e., the ratios of the mean squares for any individual source of variation compared with that for residual error. It will be noted that variance ratios were calculated only for certain selected sources of variation. That for regression, for example, is obviously so high that there is no need to quantify its statistical significance.

The value for parallelism is less than 1.00 and so there is no suggestion that the two response lines are not parallel. The value for curvature, however, is relatively high at 19.42. Reference to the appropriate statistical tables, the variance ratio tables for the F test gives values for 1 and 20 d.f. (1 d.f. for curvature and 20 d.f. for error) of 4.4 at the 5% probability level and 8.1 at the 1% probability level. Thus, a variance ratio of 19.42 in this assay indicates that curvature is highly significant. In fact the high significance is because this is a very good assay with low residual error! Sample and standard potencies are very close and there is no cause to doubt the validity of the assay.

For more information on the F test the reader may consult the publication of Hewitt (1977) or standard textbooks on statistics.

8.3   Dose–Response Lines: Type 2

Although not so common in biological work as is the type 1 response, the type 2 direct relationship between response and dose is analogous to the relationship that is so common in analytical chemistry. Thus, optical absorbance and optical rotation are directly proportional to concentration of solute; volume of titrant is directly proportional to weight of substance titrated, etc. Also, as in the cases of some chemical assays, there is sometimes a blank value to be subtracted; i.e., there is a small response even in the absence of an added dose of standard or sample.

A point of fundamental difference, however, is that in the microbiological assay it is always necessary to compare responses to sample and reference standard at the same time, whereas in the chemical assay it may be possible to compare sample response with published data or with a response to a standard that was obtained on a previous occasion.

A point of practical difference is that the microbiological test is normally carried out using several dose levels of both reference standard and sample plus a blank determination. The chemical assay frequently uses only one level.

When the observations made in a microbiological assay of this type are expressed graphically, responses to the reference standard are plotted against their actual dose. Responses to the sample of unknown potency must be plotted against nominal potencies. Unless these nominal potencies happen to coincide exactly with the true potencies, they should give lines which have a common origin corresponding to zero dose, which may or may not also coincide with zero response. In the ideal case these lines are straight and the ratio of their slopes corresponds to the ratio of the potencies of the preparations, hence the description "slope ratio assays." Examples of these ideal cases were given in Chapter 6. Straight lines coinciding at both zero dose and zero response are shown in Fig. 6.17 and coinciding at zero dose only in Fig. 6.18. In the case of the turbidimetric assays in which logarithm of response is plotted against dose, the lines slope downward from left to right as increasing doses result in lower turbidity.

8.3.1   Designs for Type 2 Assays

In contrast to type 1 assays, doses should always form an arithmetic progression, because this facilitates the calculation of potency estimates and statistical evaluation of results. However, also in contrast to the case of type 1 assays, this is a statement of the ideal. The ideal is only attained when curvature of the response lines is very slight. The reasons for this will be discussed in Section 8.3.2. Very often, however, it seems that curvature is more than slight. Some of the terms that were used in the description of designs for type 1 assays have slightly different meanings here.

Symmetry. A slope ratio assay is said to be symmetric when the number of dose levels is the same for each preparation and the number of responses to corresponding treatments are the same.

Replication. This has the same meaning as before. A symmetric assay having two dose levels plus the zero-dose control is known as the *five-point common zero* assay. Similarly, a three-dose level plus zero-dose control design would be described as a *seven-point common zero* assay. These may also be referred to as five-point assays and seven-point assays, respectively. Replication is commonly three or four tubes per treatment. A single assay may include only two preparations (reference standard and a sample), or it may include two or more samples to be compared with the standard. Although there is no theoretical limit as to the number of samples that may be compared with a standard, there are practical limits because it is highly desirable that all test solutions within a single assay be represented in replicate test tube racks. For example, three samples and one standard assayed at three dose levels plus the zero-dose control would amount to a total of 13 treatments. If replication were three, then this would entail 39 tubes to which should be added perhaps another 3 of high-dose standard purely for the purpose of checking progress of growth as the end point is approached. (These tubes, after removal from the incubator bath for reading, would not be used to provide data for the computation of potency estimate.)

Thus, for example, a rack that would accommodate 45 tubes would be suitable for such an assay and would have three unused spaces. Randomization of the tubes within the rack would minimize any bias that might otherwise arise from slight differences in temperature at different locations within the incubator bath.

8.3.2   Curvature of Type 2 Dose–Response Lines

The calculation procedure that was shown in Section 6.5.5 of Chapter 6 was based on the assumption of straight lines intersecting at zero dose. In contrast to type 1 assays, straightness of the response line remains an absolute requirement. Naturally, a slight deviation from linearity may be expected to lead to only a slight bias in the estimated potency. The problem is in quantification of what is "slight" and therefore acceptable curvature. The statistical evaluation leads to a grading of the *significance* of curvature (e.g., significant at the 95% probability level). This, however, only tells us the probability of the curvature being *real* in consideration of the extent of random error. It does not give any indication whatsoever of the absolute magnitude of that curvature or how the potency estimate will be biased if calculations are used that assume that the line is straight. It was shown by Hewitt (1977) in the case of an assay for nicotinic acid in which the curvature of the standard dose–response line could, subjectively, be described as "slight," that the following biases would arise from use of the slope ratio calculation. It seems clear that the slope ratio calculation procedure should be used only when it has been demonstrated that curvature is sufficiently slight that bias will be minimal.

True potency ratio	Potency ratio found	Bias (%)
0.50	0.524	+4.8
1.00	1.000	0.0
1.50	1.348	−10.1

8.3.3   Design and Evaluation of Type 2 Assays

Although the principles of assay design and evaluation are similar to those for type 1 assays, there are many differences in detail. The AOV leads to a value for residual error mean squares s^2. Variance ratios lead to an indication of the statistical significance of the source of variation, and three expressions analogous to Eqs. (8.1), (8.2), and (8.3) lead to the fiducial limits of the potency estimate.

The calculations are more complex than those for parallel-line assays and are not described here. However, the reader who wishes to learn more of them may refer to the publication of Hewitt (1977) (which includes some relatively simple examples) or to the works of Bliss (1950) or Finney (1964) for a fuller account.

Reference is now made to the assay of nicotinic acid which was presented as a five-dose level multiple-slope ratio assay in Example 6.2. The results only of a statistical evaluation are given here with some comments. A partial AOV is summarized in tabular form below.The deviations from regression could be broken down into its component parts, which would include quadratic, cubic, and quartic curvature, each with 1 degree of freedom, as well as deviations from these with a total of 6 degrees of freedom. Comparison of mean squares for the pooled deviations from regression with mean squares for residual error shows a variance ratio of 3.04. Reference to tables shows that for 9 and 30 degrees of freedom, respectively, this is significant at the 5% level. This tells us in statistical terms that the lines are curved, a fact that is seen readily from Fig. 6.21.

Source of variation	Degrees of freedom	Squares	Mean squares	Variance ratio
Regression	3	357,628.8		
Preparations	2	1,340.6		
Deviations from regression (by difference)	9	1,049.8	116.6	3.04
Treatments	14	360,019.2		
Residual error (by difference)	30	1,150.7	38.4	1.00
Total	44	361,169.8		

Confidence limits calculated for the two preparations are given for

Preparation 1:   0.922 to 0.964   ($p = 0.95$)

Preparation 2:   0.932 to 0.974   ($p = 0.95$)

These correspond to about $\pm 2.2\%$ of the estimated potency. In Section 8.3.2 attention was drawn to the bias that can arise through curvature of the response lines when the slope ratio calculation is used. In this case, because sample potencies are quite close to that of the reference standard, bias due to curvature will be quite small; this fact is demonstrated in Chapter 6 by Table 6.13, where results of the slope ratio calculations are compared with potencies estimated by interpolation from the standard curve.

8.4   Selection of Experimental Design

Potencies are estimated for a variety of purposes and the required precision may vary substantially.

It can be shown both for parallel-line and slope ratio assays that the width of confidence limits decreases approximately in proportion to the reciprocal of the square root of the number of observed responses to the standard plus the sample. It follows that high-precision assays are more laborious and therefore more expensive than assays of moderate or low precision.

Some pharmacopoeias define the minimum acceptable precision for antibiotic assays. For example, the British Pharmacopoeia (1980) states in several antibiotic monographs: "The precision of the assay is such that the fiducial limits of error are not less than 95 per cent and not more than 105 per cent of the estimated potency." This is the legal requirement for minimum precision. However, the British Pharmacopoeia also warns that the minimum is not always good enough! This is made clear by the following quotation from the British Pharmacopoeia (Appendix XIV B, p. A126). (Reproduced here by kind permission of the Medicines Commission of the United Kingdom.)

Guidance to manufacturers
    The required minimum precision for an acceptable assay of any particular antibiotic or preparation is defined in the appropriate monograph in the paragraph on the Assay. This degree of precision is the minimum acceptable for determining that the final product complies with the official requirements and may be inadequate for those deciding, for example, the potency which should be stated on the label or used as the basis for calculating the quantity of an antibiotic to be incorporated in a preparation. In such circumstances, assays of greater precision may be desirable, with, for instance, fiducial limits of error of the order of 98 to 102 percent. With this degree of precision, the lower fiducial limit lies close to the estimated potency. By using this limit, instead of the estimated potency, to assign a potency to the antibiotic either for labelling or for calculating the quantity to be included in a preparation, there is less likelihood of the final preparation subsequently failing to comply with the official requirements for potency.

Thus, the pharmacopoeia *defines* the minimum precision to be attained by a laboratory having a regulatory function for a public authority and *recommends* a more stringent degree of precision to be attained by the control laboratory of a manufacturing organization for release of products for sale. There are, of course, other situations to be considered. It is clear that in the establishment of a national standard by calibration against the international standard, or in the setting up of a company working standard, a very high degree of confidence should be the target; otherwise all subsequent assays for which these are used as standards will be subject to a bias of unknown direction and undesirable magnitude.

In the conduct of stability tests on pharmaceutical products, a sample may be examined repeatedly over a period of several months or perhaps a few years. It is necessary that any assay procedure used for the monitoring of such a stability test be of such precision as to detect any drift in potency over a few months. Fiducial limits of the order of $98-102\%$ ($p = 0.95$) would be desirable for this purpose.

By way of contrast, there may be circumstances in which a "screening test" is applied to a large number of samples in order to segregate the high from the low. For example, in the development of cultures for antibiotic fermentation, such a screening program might be used to select the top 10–20% of cultures, which could then be reexamined using an assay of higher precision. Conversely, a regulatory laboratory might find it convenient to examine a large number of samples of the same antibiotic, first segregating those of apparently unacceptably low potency, then retesting to obtain a potency estimate with a degree of precision meeting official requirements.

The factors to be considered in selecting an appropriate experimental design include the required precision of the potency estimate, expected potency or how closely can it be forecast for the purpose of preparing test solutions at appropriate dose levels, and the number of samples of the same active substance to be examined at the same time. In all cases work should be planned to make most efficient use of materials, facilities, and personnel. Multiple assays in which two or more samples are compared with the standard within one assay unit (plate or rack) always warrant consideration when there is more than one sample to be assayed.

A selection of assay designs is now summarized in Tables 8.2–8.5 for small plates, large plates, tubes for parallel-line assays, and tubes for slope ratio assays, respectively. In the case of plate assays, possible design is restricted by both size and shape of the plate, and so parameters such as number of dose levels, number of preparations, and degree of replication can be defined. For tube assays, however, much more flexibility is possible, and so Tables 8.4 and 8.5 present some designs only in outline.

Considering now the agar diffusion assay, assay plates should always be used to their full capacity; for example, it is inefficient and uneconomical to produce only four zones on a petri dish that can accommodate six. If only one sample is to be assayed using petri dishes, then a 3 + 3 design would make most efficient use of each dish. It would not be so efficient to use a 2 + 2 design and to duplicate the treatments of the sample in order to utilize space for all six zones, since this would result in an asymmetric assay.

To summarize, the basic design preferred by the writer for routine work is the 2 + 2 assay employing a 2 : 1 dose ratio because it is the least laborious, and it is least likely to bias due to curvature of the response line. Although some authorities insist on more than two dose levels so as to demonstrate the absence or nonsignificance of curvature, it should surely not be necessary to demonstrate this every time in a well-established assay procedure.

It is not possible to make recommendations with such confidence for tube assays of either type. Bearing in mind the generalization that more dose levels mean more work and, mathematically, are less efficient, then for parallel-line assays a good policy would be not to use a 5 + 5 if a 4 + 4 is satisfactory; and for slope ratio assays, not to use a seven-point common zero if a five-point common zero is satisfactory.

Table 8.2

Designs for Agar Diffusion Assays Using Small Plates (Petri Dishes)

Number of samples	Design designation	Zones per plate	Suitable dose ratios	Suggested replication (zones per treatment)	Comment
1	2 + 2	4	2:1, 4:1	6–20	Makes inefficient use of the plate
2	2 + 2	6	2:1, 4:1	6–20	Appropriate for two samples
1	3 + 3	6	2:1	6–20	Preferred design when only one sample is to be assayed
n	5 + 1	6	5:4	9	This is one of the U.S. CFR methods and for each sample employs three plates each having three sample zones; the design is little used in Europe

Table 8.3

Designs for Agar Diffusion Assays Using Large Square Plates

Number of samples	Design designation[a]	Zones per plate	Suitable dose ratios	Replication within plate (zones per treatment)	Comment
1	3 + 3 with 6 × 6 LS	36	2:1	6	A good design especially when good forecast of potency can be made
1	2 + 2 with 8 × 8 LS	64	2:1, 4:1	16	Each treatment twice in each row and column; 2:1 dose ratio preferred
1	4 + 4 with 8 × 8 LS	64	3:2	8	Narrow dose ratio necessary to avoid high overall dose range; dilutions inconvenient
1	HMML with 8 × 8 LS	64	2:1	High and low, 8 Medium, 16	Designed to make full use of plate, avoids difficult dilutions and wide overall range

Table 8.3 (*Continued*)

Number of samples	Design designation[a]	Zones per plate	Suitable dose ratios	Replication within plate (zones per treatment)	Comment
2	2 + 2 with 6 × 6 LS	36	2 : 1, 4 : 1	6	A good design if plate will accommodate maximum of 36 zones
2	3 + 3 with 9 × 9 LS	81	2 : 1	9	A good design, although plate is rather large and cumbersome
3	2 + 2 with 8 × 8 LS	64	2 : 1, 4 : 1	8	A good design for three samples or multiples of three; moderate precision from one plate
7	2 + 2 with 8 × 8 QLS	64	2 : 1, 4 : 1	4	Suitable when there are many samples to assay and moderate to low precision is acceptable
1	3 + 3 with 12 × 12 LS	144	2 : 1	12	Duplicate weighings of standard and sample lead to 12 treatments, 72 zones for each preparation; used for highest precision assays
2	2 + 2 with 12 × 12 LS	144	2 : 1, 4 : 1	12	Duplicate weighings of standard and sample lead to 12 treatments, 48 zones for each preparation; used for very high precision assays

(continued)

Table 8.3 (*Continued*)

Number of samples	Design designation[a]	Zones per plate	Suitable dose ratios	Replication within plate (zones per treatment)	Comment
12	4 + 1 with 8 × 8 QLS[b]	64	2 : 1	4	This incorporates a four-dose level standard curve and a single dose level for each sample; used only for low-precision work and especially when potency forecasts are weak

[a] LS, Latin square; QLS, quasi-Latin square.
[b] In place of QLS, a completely random distribution of 16 numbers has been recommended.

Table 8.4

Designs for Tube Assays of the Parallel-Line Type

Design designation	Suitable dose ratio	Corresponding dose range	Comment
4 + 4	3 : 2	3.75 : 1	This is the simplest of designs described in this group and so is the preferred design if the slope of the response line (using transformation of response if necessary) is such that an overall dose range of 3.75 : 1 is practicable.
4 + 4	$2^{0.5} : 1$	2.83 : 1	The narrower overall dose range makes this design sometimes more appropriate than the first one above; however, the dilutions are less convenient.
5 + 5	5 : 4	2.44 : 1	Dilutions for both these dose ratios are not convenient; choice between the two dose ratios must be in consideration of the slope of the response line.
5 + 5	6 : 5	2.07 : 1	
6 + 6	5 : 4	3.05 : 1	Dilutions for all three of these dose ratios are inconvenient; the use of so many dose levels has nothing to commend it other than that it may be necessary to have an excessive number because responses at one or other of the extremes are likely to be outside the usable range.
6 + 6	6 : 5	2.49 : 1	
6 + 6	10 : 9	1.69 : 1	
5 + 1	Varies	Varies	This is one of the designs described by the U.S. CFR; methods for calculation of potency estimate are described but interpolation from the standard curve may be more appropriate; it has the inherent disadvantages of an asymmetric assay.

Table 8.5

Designs for Tube Assays of the Slope Ratio Type

Design designation	Comment
Three-point common zero	This is the simplest possible form of the slope ratio assay and is also the most efficient, since no effort is expended in the preparation of intermediate-dose test solutions which, in any event, contribute less "weight" to the estimation of potency. This is the form that is, effectively, in common use in chemical and physical assay methods even though not generally known by the name "slope ratio assay." However, this simple design includes no checks for curvature and so it is not recommended for general use in microbiological assay.
Five-point common zero	This is the simplest slope ratio assay that incorporates a check for curvature. Provided that assay conditions can be so regulated that the dose range covers the greater part of the range of linear response, then this is the design of choice. Modification of the design to include additional samples for comparison with the single standard is desirable if more than one sample is to be examined.
Seven-point common zero	This assay is inherently less efficient than the five-point assay; however, it does offer the possibility of extending the dose range further. In the event that the higher doses yield responses outside the range of linearity, then these may be discarded and the remaining data processed as a five-point assay.

References

Anon. (1972). EEC Commission Directive 72/199.
Anon. (1985). "British Animal Feeding Stuffs Regulations," SI No. 273.
Bliss, C. I. (1950). *In* "Biological Standardization" (J. H. Burn, *et al.,* eds.). Oxford University Press, Oxford.
British Pharmacopoeia (1980). British Pharmacopoeia Commission, H. M. Stationary Office, London.
Burn, J. H., *et al.,* eds. (1950) "Biological Standardization." Oxford University Press, Oxford.
Emmens, C. W. (1948). "Principles of Biological Assay." Chapman and Hall, London.
European Pharmacopœia (1980). European Pharmacopœia Commission, Strasbourg.
Finney, D. J. (1964). "Statistical Method in Biological Assay." Griffin, London.
Hewitt, W. (1977). "Microbiological Assay." Academic Press, New York and London.
Hewitt, W. (1981). *J. Biol. Standardization* **9**, 1.
International Pharmacopœia (second edition) (1967). World Health Organization, Geneva.
Kavanagh, F. W. (1968). *Appl. Microbiol.* **16**, 777.
Vincent, S. (1983). Unpublished data.
Vincent, S. (1986). Unpublished data.

CHAPTER 9

DIFFUSION ASSAY METHODS FOR ANTIBIOTICS

9.1 Introduction

It is not possible — or even intended — to present an assay outline for every available growth-inhibiting substance in the following pages. In Section 9A of this chapter examples of some well-tested outline assay methods are given which the authors have found to be most satisfactory for routine work, utilizing $2 + 2$ or $3 + 3$ Latin square designs. In some cases more than one method is described for the same growth-inhibiting substance, in which case an explanation is provided to indicate under which circumstances the alternative method could be useful. A few examples are also included for the assay of antibiotic mixtures. For further information relevant to the assay of mixed antibiotics, the reader is referred to the tables of Arret et al. (1957), who provide data on interference threshold of a second antibiotic in the assay of the first, and to Weiss et al. (1957) and Andrew and Weiss (1959), who give data on solubilities of antibiotics in a range of solvents, which could permit their physical separation. Tables of both interference thresholds and solubilities based on these publications were reproduced by Kavanagh (1963).

The outline methods for individual assays presented on the following pages must be supplemented by the more detailed guidelines given in the previous chapters. Formulas of assay media and diluents follow the numbering system adopted by 21 CFR (U.S. Code of Federal Regulations) where appropriate and are to be found in Appendix 4. Most of these formulas can also be found in several national pharmacopoeias and other publications, but are not always numbered in the same order.

Owing to the widespread use of large plates in Great Britain in preference to petri dishes, single-layer media have become the norm. It is difficult to pour a seed layer of uniform thickness using large plates, because the seed layer tends to solidify before it has a chance to spread evenly. For this reason, only one medium is given for the assays described in this section. Even though these assays were originally developed for large plates, they can be used for petri dish assays of the $2 + 2$ pattern. Similarly, the two dose levels can be easily turned into three dose levels by including one more dose level either below or above those given, provided the dose ratio is maintained.

In cases of dispute over potencies with other manufacturers or regulatory authorities, the official assay methods given in the various compendiums must be followed. The currently available official methods are summarized in Table 9.1.

Table 9.1

Microbiological Assay Methods Listed in Official Compendia[a]

Antibiotic	21CFR(1985)	USPXXI	BP1980	IP Ed.3	EP Ed.2
Amikacin	t	t	—	—	—
Amoxicillin	d	—	—	—	—
Amphomycin	d	—	—	—	—
Amphotericin B	d	d	d	—	—
Ampicillin	d	d	—	—	—
Bacitracin Zinc	d	d	d	d	—
Bleomycin	d	d	—	—	—
Candicidin	t	t	d	—	—
Capreomycin	t	t	d	—	—
Carbenicillin	d	d	—	—	—
Cefactor	d	—	—	—	—
Cefadroxil	d	—	—	—	—
Cefamandol	d	—	—	—	—
Cefazolin	d	—	—	—	—
Cefotaxime	d	—	—	—	—
Cefoxitin	d	—	—	—	—
Cephalexin	d	d	—	d	—
Cephaloglycin	d	—	—	—	—
Cephaloridine	d	—	—	—	—
Cephalothin	d	d	—	d	—
Cephapirin	d	d	—	—	—
Cephradine	d	d	—	—	—
Chloramphenicol	t	t	—	—	—
Chlortetracycline	t	t	d and t	d	d and t
Clindamycin	d	d	—	—	—
Cloxacillin	d	d	—	d	—
Colistimethate Sodium	d	d	d	—	—
Colistin	d	d	d	—	—
Cyclacillin	d	—	—	—	—
Cycloserine	t	t	—	—	—
Dactinomycin	d	d	—	—	—
Demeclocycline	t	t	d and t	—	d and t
Dicloxacillin	d	d	—	d	—
Dihydrostreptomycin	d and t	d and t	—	—	—
Doxycycline	t	t	d	—	—
Erythromycine	d	d	d and t	d	d and t
Framycetin	—	—	d	—	—
Gentamicin	d	d	d	—	—

(continued)

Table 9.1. (*Continued*)

Antibiotic	21CFR(1985)	USPXXI	BP1980	IP Ed.3	EP Ed.2
Gramicidin	t	t	—	—	—
Kanamycin	t	t	d	—	d
Kanamycin B	d	—	—	—	—
Lymecycline	—	—	d	—	—
Lincomycin	t	t	—	—	—
Meclocycline	t	—	—	—	—
Methacycline	t	t	—	—	—
Methicillin	d	d	—	—	—
Minocycline	t	t	—	—	—
Mitomycin	d	d	—	—	—
Nafcillin	d	d	—	—	—
Natamycin	d	d	—	—	—
Neomycin	d	d	d and t	d	d and t
Netilmicin	d	d	—	—	—
Novobiocin	d	d	—	d	—
Nystatin	d	d	d	d	—
Oleandomycin	d	—	—	—	—
Oxacillin	d	d	—	d	—
Oxytetracycline	t	t	d and t	d	d and t
Paromomycin	d	d	—	—	—
Penicillin G	d	d	—	—	—
Penicillin V Potassium	d	—	—	—	—
Plicamycin	d	d	—	—	—
Polymyxin B	d	d	d	d	d
Rifampin	d	d	—	—	—
Rifamycin	—	—	d and t	—	d and t
Rolitetracycline	t	t	—	—	—
Sisomicin	d	d	—	—	—
Spectinomycin	t	t	—	—	—
Streptomycin	d and t	t	d and t	d	d and t
Tetracycline	t	t	d and t	d	d and t
Ticarcillin	d	d	—	—	—
Tobramycin	t	t	d and t	—	—
Troleandomycin	t	t	—	—	—
Tyrothricin	t	—	—	—	—
Vancomycin	d	d	d	—	—
Viomycin	—	t	—	—	—
TOTAL	73	57	32	15	18
DIFFUSION ASSAYS	50	35	23	15	10
TUBE ASSAYS	23	22	9	0	8

[a] d, Agar diffusion assays; t, tube assays; —, not listed; CFR, Code of Federal Regulations (US); USP, United States Pharmacopoeia; BP, British Pharmacopoeia; IP, International Pharmacopoeia; EP, European Pharmacopoeia; Ed., Edition.

Section 9B describes the cylinder–petri dish assay methods listed in 21 CFR (1985) in an easy-to-follow manner. The CFR methods employ a 5 + 1 design (see Section 8.2.1), base and seed layers in petri dishes and cylinders (see Section 4.6.3). 21 CFR contains the largest number of official antibiotic diffusion assay methods, as shown in Table 9.1. However, their presentation here does not imply either recommendation or approval of the 5 + 1 statistical design advocated by CFR. By selecting appropriate dose levels, any of the assays presented in Section 9B can be easily converted to a 2 + 2 or 3 + 3 Latin square pattern.

9A ROUTINE METHODS

9.2 Amikacin

Organism: *Bacillus pumilus* NCTC 8241
Prepare a concentrated spore suspension as described in Section 2.8.
Inoculum: Mix about 1.0% (v/v) of the calibrated spore suspension to the assay medium held at about 60°C.
Medium: Antibiotic medium 5. Final pH before inoculation adjusted to pH 7.9.
Diluents: (1) Distilled water. (2) Phosphate buffer, pH 8.0 (solution 3).
Standard: A working standard of amikacin sulfate of known chemical purity, e.g., 134 mg amikacin sulfate are equivalent to 100 mg of amikacin base.
Dose levels: 2.0 μg/ml and 1.0 μg/ml (dose ratio 2 : 1)
Accurately weigh about 67 mg amikacin sulfate working standard and quantitatively transfer to a 50-ml volumetric flask. Dissolve in distilled water and make volume up to 50 ml with distilled water. Prepare all subsequent dilutions with phosphate buffer, pH 8.0. For example:

$$67 \pm 1 \text{ mg} \rightarrow 50 \text{ ml} : 10 \text{ ml} \rightarrow 100 \text{ ml} : 10 \text{ ml} \rightarrow 50 \text{ ml}$$
$$: 10 \text{ ml} \rightarrow 100 \text{ ml}$$

Store the primary stock solution of the working standard at 4°C for not longer than 14 days.
Incubation: 37°C overnight
Remarks: This method is also suitable for the assay of amikacin in the presence of cefuroxime, up to a ratio of 1 : 5.5 of the former to the latter.

9.3 Amoxicillin

Organism: *Bacillus subtilis* ATCC 6633, NCIB 8533
Prepare a concentrated spore suspension as detailed in Section 2.8.
Inoculum: Mix about 1.0% (v/v) of the calibrated spore suspension to the assay medium held at about 60°C.
Medium: Antibiotic medium 40. Final pH before inoculation adjusted to pH 7.2.

Diluents: (1) 0.1 *N* Hydrochloric acid. (2) Distilled water. (3) Phosphate buffer, pH 7.5 (solution 20).

Standard: A working standard of amoxicillin trihydrate of known chemical purity. Express the results in terms of the anhydrous amoxicillin.

Dose levels: 2.0 μg/ml and 1.0 μg/ml (dose ratio 2:1)

Accurately weigh about 117 mg amoxicillin working standard and quantitatively transfer into a 50-ml volumetric flask. Dissolve in 10 ml 0.1 *N* hydrochloric acid and make volume up to 50 ml using distilled water. Prepare subsequent dilutions with phosphate buffer, pH 7.5. For example:

$$117 \pm 1 \text{ mg} \rightarrow 50 \text{ ml} : 5 \text{ ml} \rightarrow 500 \text{ ml} : 5 \text{ ml} \rightarrow 50 \text{ ml}$$
$$: 5 \text{ ml} \rightarrow 100 \text{ ml}$$

Store the primary stock solution of the working standard at 4°C for not longer than 7 days.

Incubation: 30°C overnight

9.4 Ampicillin

Organism: *Bacillus subtilis* ATCC 6633, NCIB 8533

Prepare a concentrated spore suspension as detailed in Section 2.8.

Inoculum: Mix about 1.0% (v/v) of the calibrated spore suspension to the assay medium held at about 60°C.

Medium: Antibiotic medium 2. Final pH before inoculation adjusted to pH 7.5.

Diluents: (1) Distilled water. (2) Phosphate buffer, pH 7.5 (solution 20).

Standard: A working standard of ampicillin of known chemical purity. Express the results in terms of the anhydrous ampicillin.

Dose levels: 1.0 μg/ml and 0.5 μg/ml (dose ratio 2:1)

Accurately weigh sufficient working standard to prepare a primary stock solution of 500 μg/ml, calculated on the anhydrous basis. For example, accurately weigh about 60 mg of ampicillin trihydrate of 82.5% pure and dissolve it in 100 ml distilled water. Further dilute in pH 7.5 phosphate buffer to plating levels.

$$60 \pm 1 \text{ mg} \rightarrow 100 \text{ ml} : 5 \text{ ml} \rightarrow 250 \text{ ml} : 5 \text{ ml} \rightarrow 50 \text{ ml}$$
$$: 5 \text{ ml} \rightarrow 100 \text{ ml}$$

Store the primary stock solution of the working standard at 4°C for not longer than 7 days.

Incubation: 30°C overnight

9.5 Azlocillin

Organism: *Bacillus subtilis* ATCC 6633, NCIB 8533

Prepare a concentrated spore suspension as detailed in Section 2.8.

Inoculum: Mix about 1.0% (v/v) of the calibrated spore suspension to the assay medium held at about 60°C.

Medium: Antibiotic medium 2. Final pH before inoculation adjusted to pH 7.5.

Diluent: Phosphate buffer, pH 7.5 (solution 20)

Standard: A working standard of known chemical purity

Dose levels: 4 μg/ml and 2 μg/ml (dose ratio 2:1)

Accurately weigh about 100 mg of azlocillin working standard, and dissolve it in 500 ml phosphate buffer, pH 7.5, to prepare a primary stock solution of 200 μg/ml. Further dilute in pH 7.5 phosphate buffer as shown:

$$100 \pm 1 \text{ mg} \rightarrow 500 \text{ ml} : 10 \text{ ml} \rightarrow 100 \text{ ml} : 10 \text{ ml} \rightarrow 50 \text{ ml}$$
$$: 10 \text{ ml} \rightarrow 100 \text{ ml}$$

Store the primary stock solution of the working standard at 4°C for not longer than 7 days.

Incubation: 30°C overnight

Remarks: This method is also suitable for the assay of azlocillin even in the presence of equal concentration of cefuroxime.

9.6   Bacitracin

Organism: *Micrococcus luteus* ATCC 10240b, NCIB 8994

Maintain the test organism by regular subculturing on slants of the assay medium containing about 125 IU/ml neomycin and incubating at 30°C for 48 hours.

Inoculum: Suspend the growth from three freshly grown slants in 15 ml of sterile saline, and mix about 1.0% (v/v) of the calibrated suspension to the assay medium held at 50°C. The inoculum is usable for up to 14 days if stored at 4°C.

Medium: Antibiotic medium 1. Final pH before inoculation adjusted to pH 7.5.

Diluents: (1) Distilled water. (2) 3 N Hydrochloric acid (HCl). (3) Phosphate buffer, pH 7.5 (solution 20).

Standard: A working standard of zinc bacitracin calibrated directly or indirectly against the international biological standard.

Dose levels: 1.0 IU/ml and 0.5 IU/ml (dose ratio 2 : 1)

For example, after drying at a vacuum of 5 mm of mercury and a temperature of 60°C for 3 hours, accurately weigh about 90 mg of the working standard having a potency of 55.8 IU/mg and transfer quantitatively to a 100-ml volumetric flask. Add about 10 ml of distilled water and 1.0 ml of 3 N HCl; leave it to stand at room temperature for 30 minutes, then dilute to volume with distilled water to produce a primary stock solution of 50 IU/ml. Further dilute in pH 7.5 phosphate buffer to plating-out levels:

$$90 \pm 1 \text{ mg} \rightarrow 100 \text{ ml} : 5 \text{ ml} \rightarrow 250 \text{ ml}$$
$$: 5 \text{ ml} \rightarrow 500 \text{ ml}$$

Store the primary stock solution of the working standard at 4°C for not longer than 7 days. Dilutions of samples must contain the same amount of HCl as the standard solutions.

Incubation: 37°C overnight

Remarks: This method is suitable for the assay of bacitracin either on its own or in the presence of neomycin. The test organism must be maintained on slants of assay medium containing neomycin to preserve its properties of resistance to neomycin while maintaining its sensitivity to bacitracin. If the test organism is maintained in the absence of neomycin, there is a danger that it might revert to a nonresistant form and become sensitive to both antibiotics.

9.7   Benzalkonium Chloride

Organism: *Micrococcus luteus* ATCC 10240, NCIB 8166

Maintain the test organism by regular subculturing on slants of assay medium and incubating at 30°C for 48 hours.

Inoculum: Suspend the growth from a freshly grown slant in about 10 ml of sterile saline, and mix about 1.0% (v/v) of the calibrated suspension to the assay medium held at 50°C. The inoculum is usable for up to 2 weeks if stored refrigerated at about 4°C.

Medium: Antibiotic medium 1. Final pH before inoculation adjusted to pH 7.5.

Diluent: Distilled water

Standard: A working standard of pure benzalkonium chloride (BKC) of known chemical purity. For routine purposes the commercially available 50 ± 1% solution is acceptable.

Dose levels: 0.005% (w/v) and 0.0005% (w/v) (dose ratio 10 : 1)

Accurately weigh sufficient working standard and dissolve in distilled water to prepare a 0.5% (w/v) stock solution. Further dilute in distilled water to plating-out levels. For example, 500 mg pure BKC → 100 ml or

$$5.0 \text{ g } (50\%, \text{w/v}) \rightarrow 500 \text{ ml}: 10 \text{ ml} \rightarrow 100 \text{ ml}: 5 \text{ ml} \rightarrow 50 \text{ ml}$$
$$: 5 \text{ ml} \rightarrow 500 \text{ ml}$$

The aqueous solution foams strongly when shaken. Store the primary stock solution of the working standard at 4°C for not longer than 1 month.

Incubation: 30°C overnight

Remarks: Preservatives are assayed by the same general techniques as antibiotics. For further references see Vincent and Shaw (1972).

9.8   Carbenicillin

Organism: *Bacillus subtilis* ATCC 6633, NCIB 8533

Prepare a concentrated spore suspension as detailed in Section 2.8.

Inoculum: Mix about 1.0% (v/v) of the calibrated spore suspension to the assay medium held at about 60°C.

Medium: Antibiotic medium 41. Final pH before inoculation adjusted to pH 7.0.

Diluent: Phosphate buffer pH 7.5 (solution 20)

Standard: A working standard of known chemical purity. Express the results in terms of the anhydrous compound.

Dose levels: 4.0 μg/ml and 2.0 μg/ml (dose ratio 2 : 1)

Accurately weigh about 90 ± 1 mg of the working standard and dissolve quantitatively in 200 ml of phosphate buffer, pH 7.5. Further dilute in pH 7.5 phosphate buffer to plating-out levels. For example:

$$90 \pm 1 \text{ mg} \rightarrow 200 \text{ ml}: 10 \text{ ml} \rightarrow 100 \text{ ml}: 10 \text{ ml} \rightarrow 100 \text{ ml}$$
$$: 10 \text{ ml} \rightarrow 200 \text{ ml}$$

Store the primary stock solution of the working standard at 4°C for not longer than 14 days.

Incubation: 30°C overnight

9.9   Cefazolin

Organism: *Bacillus subtilis* ATCC 6633, NCIB 8533

Prepare a concentrated spore suspension as detailed in Section 2.8.

Inoculum: Mix about 1.0% (v/v) of the calibrated spore suspension to the assay medium held at about 60°C.

Medium: Antibiotic medium 42. Final pH before inoculation adjusted to pH 6.5.

Diluents: (1) Phosphate buffer, pH 8.0 (solution 3). (2) Distilled water.
Standard: A working standard of known chemical purity. Express the results in terms of the anhydrous compound.
Dose levels: 1.0 μg/ml and 0.5 μg/ml (dose ratio 2:1)
Accurately weigh about 80 mg of the working standard and dissolve quantitatively in 200 ml of phosphate buffer, pH 8.0. Further dilute in distilled water to plating-out levels. For example:

$$80 \pm 1 \text{ mg} \rightarrow 200 \text{ ml}: 5 \text{ ml} \rightarrow 100 \text{ ml}: 5 \text{ ml} \rightarrow 100 \text{ ml}$$
$$: 5 \text{ ml} \rightarrow 200 \text{ ml}$$

Store the primary stock solution of the working standard at 4°C for not longer than 5 days.
Incubation: 37°C overnight

9.10 Cefotaxime

Organism: *Bacillus subtilis,* not deposited with ATCC or NCIB; Glaxo Group Research Culture Collection No. 1768E
Prepare a concentrated spore suspension as detailed in Section 2.8.
Inoculum: Mix about 1.0% (v/v) of the calibrated spore suspension to the assay medium held at about 60°C.
Medium: Antibiotic medium 42. Final pH before inoculation adjusted to pH 7.0.
Diluent: Phosphate buffer, pH 4.5 (solution 4)
Standard: A working standard of known chemical purity. Express the results in terms of the anhydrous compound.
Dose levels: 0.2 μg/ml and 0.1 μg/ml (dose ratio 2:1)
Accurately weigh about 80 mg of the working standard and dissolve quantitatively in 200 ml of phosphate buffer, pH 4.5. Further dilute in pH 4.5 phosphate buffer to plating-out levels. For example:

$$80 \pm 1 \text{ mg} \rightarrow 200 \text{ ml}: 5 \text{ ml} \rightarrow 100 \text{ ml}: 10 \text{ ml} \rightarrow 100 \text{ ml}: 10 \text{ ml} \rightarrow 100 \text{ ml}$$
$$: 10 \text{ ml} \rightarrow 200 \text{ ml}$$

Use the standard solution the same day and do not store.
Incubation: 37°C overnight

9.11 Ceftazidime

Organism: *Bacillus subtilis,* not deposited with ATCC or NCIB; Glaxo Group Research Culture Collection No. 1768E
Prepare a concentrated spore suspension as detailed in Section 2.8.
Inoculum: Mix about 1.0% (v/v) of the calibrated spore suspension to the assay medium held at about 60°C.
Medium: Antibiotic medium 43. Final pH before inoculation adjusted to pH 6.0.
Diluent: Phosphate buffer, pH 4.5 (solution 4)
Standard: A working standard of known chemical purity. The pentahydrate is usually mixed with an inert solubilizing agent (e.g., sodium carbonate) at the rate of 10% (w/w).
Dose levels: 2.0 μg/ml and 1.0 μg/ml (dose ratio 2:1)
Accurately weigh about 116 mg (equivalent to ∼ 100 mg of the pentahydrate) of the working

standard and dissolve quantitatively in 500 ml of phosphate buffer, pH 4.5. Further dilute in pH 4.5 phosphate buffer to plating-out levels. For example:

$$116 \pm 1 \text{ mg} \rightarrow 500 \text{ ml}: 10 \text{ ml} \rightarrow 100 \text{ ml}: 10 \text{ ml} \rightarrow 100 \text{ ml}$$
$$: 10 \text{ ml} \rightarrow 200 \text{ ml}$$

Store the primary stock solution of the working standard at 4°C, and use it within 48 hours.
Incubation: 37°C overnight

9.12   Cefuroxime

Organism: *Bacillus subtilis,* not deposited with ATCC or NCIB; Glaxo Group Research Culture Collection No. 1768E
Prepare a concentrated spore suspension as detailed in Section 2.8.
Inoculum: Mix about 1.0% (v/v) of the calibrated spore suspension to the assay medium held at about 60°C.
Medium: Antibiotic medium 42. Final pH before inoculation adjusted to pH 6.5.
Diluents: (1) Distilled water. (2) Phosphate buffer, pH 4.5 (solution 4).
Standard: A working standard of known chemical purity
Dose levels: 0.2 μg/ml and 0.1 μg/ml (dose ratio 2 : 1)
Accurately weigh about 50 mg of the working standard and quantitatively dissolve in 50 ml of distilled water. Further dilute in pH 4.5 phosphate buffer to plating-out levels, e.g., standard potency 950 μg/mg:

$$50 \pm 1 \text{ mg} \rightarrow 50 \text{ ml}: 5 \text{ ml} \rightarrow 500 \text{ ml}: 5 \text{ ml} \rightarrow 250 \text{ ml}$$
$$: 5 \text{ ml} \rightarrow 500 \text{ ml}$$

Store the working standard stock solution at 4°C for not longer than 7 days.
Incubation: 37°C overnight

9.13   Cephalexin

Organism: *Bacillus subtilis,* not deposited with ATCC or NCIB; Glaxo Group Research Culture Collection No. 1768E
Prepare a concentrated spore suspension as detailed in Section 2.8.
Inoculum: Mix about 1.0% (v/v) of the calibrated spore suspension to the assay medium held at about 60°C.
Medium: Antibiotic assay medium 42. Final pH before inoculation adjusted to pH 7.2.
Diluents: (1) 1 N HCl. (2) Distilled water. (3) Phosphate buffer, pH 4.5 (solution 4).
Standard: A working standard of known chemical purity. Express the results in terms of the anhydrous compound.
Dose levels: 2.0 μg/ml and 1.0 μg/ml (dose ratio 2 : 1)
With a standard potency of 93.6% (w/w), accurately weigh about 105 mg of the working standard and quantitatively dissolve in 10 ml of 1 N HCl, then dilute to 100 ml with distilled water. Further dilute with pH 4.5 phosphate buffer to plating-out levels. For example:

$$105 \pm 1 \text{ mg} \rightarrow 100 \text{ ml} : 5 \text{ ml} \rightarrow 250 \text{ ml} : 10 \text{ ml} \rightarrow 100 \text{ ml}$$
$$: 10 \text{ ml} \rightarrow 200 \text{ ml}$$

Store this working standard stock solution at 4°C for not longer than 7 days.
Incubation: 30°C overnight

9.14   Cephaloridine

Organism: *Bacillus subtilis,* not deposited with ATCC or NCIB; Glaxo Group Research Culture Collection No. 1768E
Prepare a concentrated spore suspension as detailed in Section 2.8.
Inoculum: Mix about 1.0% (v/v) of the calibrated spore suspension to the assay medium held at about 60°C.
Medium: Antibiotic medium 43. Final pH before inoculation adjusted to pH 6.0.
Diluent: Distilled water
Standard: A working standard of known chemical purity. Express the results in terms of the anhydrous compound.
Dose levels: 0.05 µg/ml and 0.025 µg/ml (dose ratio 2:1)
With a standard potency of 96.8% (w/w), accurately weigh about 103 mg of the working standard and quantitatively dissolve in 100 ml of distilled water. Further dilute with distilled water to plating-out levels. For example:

$$103 \pm 1 \text{ mg} \rightarrow 100 \text{ ml} : 5 \text{ ml} \rightarrow 500 \text{ ml} : 5 \text{ ml} \rightarrow 100 \text{ ml} : 10 \text{ ml} \rightarrow 100 \text{ ml}$$
$$: 10 \text{ ml} \rightarrow 200 \text{ ml}$$

Store the working standard stock solution at 4°C for not longer than 7 days.
Incubation: 30°C overnight
Remarks: For further details refer to Shaw and Vincent (1972).

9.15   Cephalothin

Organism: *Bacillus subtilis* ATCC 6633, NCIB 8533
Prepare a concentrated spore suspension as detailed in Section 2.8.
Inoculum: Mix about 1.0% (v/v) of the calibrated spore suspension to the assay medium held at 60°C.
Medium: Antibiotic medium 2. Final pH before inoculation adjusted to pH 6.5.
Diluent: Distilled water
Standard: A working standard of known chemical composition and purity. Express the results in terms of the anhydrous compound.
Dose levels: 1.0 µg/ml and 0.5 µg/ml (dose ratio 2:1)
With a standard potency of 97.9% (w/w), accurately weigh about 103 mg of the working standard and quantitatively dissolve in 100 ml of distilled water. Further dilute the distilled water to plating-out levels. For example:

$$103 \pm 1 \text{ mg} \rightarrow 100 \text{ ml} : 5 \text{ ml} \rightarrow 500 \text{ ml} : 10 \text{ ml} \rightarrow 100 \text{ ml}$$
$$: 10 \text{ ml} \rightarrow 200 \text{ ml}$$

Store the working standard primary stock solution at 4°C for not longer than 7 days.
Incubation: 30°C overnight

9.16   Chlortetracycline

Organism: *Bacillus subtilis* ATCC 6633, NCIB 8533
Prepare a concentrated spore suspension as detailed in Section 2.8.
Inoculum: Mix about 1.0% (v/v) of the calibrated spore suspension to the assay medium held at 60°C.
Medium: Antibiotic medium 2. Final pH before inoculation adjusted to pH 6.5.
Diluents: (1) 0.01 N HCl. (2) Phosphate buffer, pH 4.5 (solution 4).
Standard: A working standard of known chemical purity, calibrated directly or indirectly against the international biological standard.
Dose levels: 0.2 IU/ml and 0.1 IU/ml
With a standard potency of 947 IU/mg, accurately weigh about 100 mg of the working standard and quantitatively dissolve in 100 ml of 0.01 N HCl. Further dilute with pH 4.5 phosphate buffer to plating-out levels. For example:

$$100 \pm 1 \text{ mg} \rightarrow 100 \text{ ml}: 5 \text{ ml} \rightarrow 500 \text{ ml}: 5 \text{ ml} \rightarrow 250 \text{ ml}$$
$$: 5 \text{ ml} \rightarrow 500 \text{ ml}$$

Store the working standard primary stock solution at 4°C for not longer than 7 days.
Incubation: 30°C overnight

9.17   Cloxacillin

Organism: *Bacillus subtilis* ATCC 6633, NCIB 8533
Prepare a concentrated spore suspension as detailed in Section 2.8.
Inoculum: Mix about 1.0% (v/v) of the calibrated spore suspension to the assay medium held at 60°C.
Medium: Antibiotic medium 41. Final pH before inoculation adjusted to pH 6.5.
Diluent: Phosphate buffer, pH 7.5 (solution 20)
Standard: A working standard of known chemical purity. Express results in terms of the anhydrous compound.
Dose levels: 5.0 μg/ml and 2.5 μg/ml (dose ratio 2:1)
Accurately weigh about 80 mg of the working standard and quantitatively dissolve in 200 ml of phosphate buffer, pH 7.5. Further dilute in pH 7.5 phosphate buffer to plating-out levels. For example:

$$80 \pm 1 \text{ mg} \rightarrow 200 \text{ ml}: 5 \text{ ml} \rightarrow 100 \text{ ml}: 25 \text{ ml} \rightarrow 100 \text{ ml}$$
$$: 25 \text{ ml} \rightarrow 200 \text{ ml}$$

Store the working standard primary stock solution at 4°C for not longer than 7 days.
Incubation: 30°C overnight

9.18   Dihydrostreptomycin

Organism: *Bacillus subtilis* ATCC 6633, NCIB 8533
Prepare a concentrated spore suspension as detailed in Section 2.8.
Inoculum: Mix about 1.0% (v/v) of the calibrated spore suspension to the assay medium held at 60°C.

Medium: Antibiotic medium 40. Final pH before inoculation adjusted to pH 7.5.

Diluent: Phosphate buffer, pH 7.5 (solution 20)

Standard: A working standard of dihydrostreptomycin sulfate of known chemical purity and potency.

Dose levels: 4.0 IU/ml and 2.0 IU/ml (dose ratio 2:1)

For example, after drying at a vacuum of 5 mm Hg and a temperature of 60°C for 3 hours, accurately weigh about 50 mg of the working standard having a potency of 772 U/mg and quantitatively dissolve in 100 ml of phosphate buffer, pH 7.5. Further dilute in pH 7.5 phosphate buffer to plating-out levels:

$$50 \pm 1 \text{ mg} \rightarrow 100 \text{ ml}: 5 \text{ ml} \rightarrow 100 \text{ ml}: 10 \text{ ml} \rightarrow 50 \text{ ml}$$
$$: 10 \text{ ml} \rightarrow 100 \text{ ml}$$

Store the working standard primary stock solution at 4°C for not longer than 1 month.

Incubation: 37°C overnight

Remarks:

1. To assay dihydrostreptomycin in the presence of penicillin, add sufficient penicillinase to the melted assay medium before inoculation to inactivate the penicillin in the sample.

2. To assay dihydrostreptomycin in the presence of streptomycin, add 10 ml of a 0.5% (w/v) aqueous semicarbazide solution to the sample solution, which was previously buffered to between pH 6.5 and 7.0, and allow to stand at room temperature for 1 hour; then dilute with phosphate buffer, pH 7.5 to plating-out levels. This method will inactivate up to 10,000 U/ml streptomycin.

9.19 Gentamicin

Organism: *Bacillus pumilus* NCTC 8241

Prepare a concentrated spore suspension as described in Section 2.8.

Inoculum: Mix about 1.0% (v/v) of the calibrated spore suspension to the assay medium held at 60°C.

Medium: Antibiotic medium 2. Final pH before inoculation adjusted to pH 7.9 (same as medium 5).

Diluent: Phosphate buffer, pH 8.0 (solution 3)

Standard: A working standard of gentamicin sulfate, calibrated directly or indirectly against the international biological standard.

Dose levels: 2.0 IU/ml and 0.5 IU/ml (dose ratio 4:1)

About 160 mg of the working standard are equivalent to 100 mg of the base. After drying at a vacuum of 5 mm Hg and a temperature of 60°C for 3 hours, accurately weigh about 160 mg of the working standard and quantitatively dissolve in 100 ml of phosphate buffer, pH 8.0. Further dilute in the same buffer to plating-out levels:

$$160 \pm 1 \text{ mg} \rightarrow 100 \text{ ml}: 10 \text{ ml} \rightarrow 500 \text{ ml}: 10 \text{ ml} \rightarrow 100 \text{ ml}$$
$$: 5 \text{ ml} \rightarrow 200 \text{ ml}$$

Store the primary stock solution of the working standard at 4°C for not longer than 1 month.

Incubation: 37°C overnight

Remarks: This method is also suitable for the assay of gentamicin in the presence of five times the amount (w/w) of cefuroxime.

9.20   Neomycin

9.20.1   Method 1

Organism: *Escherichia coli* 10/5, NCIB 10072
Maintain the test organism by regular subculturing on slants of assay medium and incubating at 37°C for 48 hours.
Inoculum: Suspend the growth from one freshly grown slant in about 10 ml sterile 0.1% (w/v) peptone diluent (solution 21), and make up volume to about 80 ml with the same solution. Mix about 1.0% (v/v) of the calibrated suspension to the assay medium held at about 48°C.
Medium: Antibiotic medium 11. Medium 1 can be used if the final pH is adjusted to pH 7.9 before inoculation.
Diluent: Phosphate buffer, pH 8.0 (solution 3)
Standard: A working standard of neomycin sulfate, calibrated directly or indirectly against the international biological standard.
Dose levels: 20 IU/ml and 5 IU/ml (dose ratio 4:1)
For example, after drying at a vacuum of 5 mm Hg and a temperature of 60°C for 3 hours, accurately weigh about 110 mg of the working standard having a potency of 758 IU/mg and dissolve quantitatively in 200 ml of phosphate buffer, pH 8.0. Further dilute with the same solution to plating-out levels:

$$110 \pm 1 \text{ mg} \rightarrow 200 \text{ ml}:5 \text{ ml} \rightarrow 100 \text{ ml}:25 \text{ ml} \rightarrow 100 \text{ ml}$$

Store the primary stock solution of the working standard at 4°C for not longer than 2 weeks.
Incubation: 37°C overnight
Remarks: This method is suitable for the assay of neomycin in the presence of bacitracin.

9.20.2   Method 2

Organism: *Bacillus pumilus* NCIB 8241
Prepare a concentrated spore suspension as described in Section 2.8.
Inoculum: Mix about 1.0% (v/v) of the calibrated spore suspension to the assay medium held at 60°C.
Medium: Antibiotic assay medium 2. Adjust final pH before inoculation to pH 8.5 (pH higher than medium 11).
Diluent: Phosphate buffer, pH 8.0 (solution 3)
Standard: As in method 1 (Section 9.20.1)
Dose levels: 0.4 IU/ml and 0.2 IU/ml (dose ratio 2:1)
Prepare the primary stock solution of the working standard as in method 1 (Section 9.20.1), and dilute with pH 8.0 phosphate buffer to plating-out levels. For example:

$$110 \pm 1 \text{ mg} \rightarrow 200 \text{ ml}:10 \text{ ml} \rightarrow 100 \text{ ml}:10 \text{ ml} \rightarrow 100 \text{ ml}:10 \text{ ml} \rightarrow 100 \text{ ml}$$
$$:10 \text{ ml} \rightarrow 200 \text{ ml}$$

Storage of standard solution as in method 1 (Section 9.20.1).
Incubation: 30°C overnight
Remarks: This method is suitable for the assay of neomycin in the presence of thiomersal.

9.20.3 Method 3

Organism: *Micrococcus luteus,* resistant to streptomycin and dihydrostreptomycin; not deposited with ATCC or NCIB; Glaxo Group Research Culture Collection No. 400E SDR

Maintain the test organism by regular subculturing on slants of the assay medium containing about 100 IU/ml of streptomycin and incubating at 37°C for 48 hours.

Inoculum: Suspend the growth from a freshly grown slant in about 50 ml of sterile saline and mix about 1.0% (v/v) of the calibrated suspension to the assay medium held at 50°C. The inoculum is usable for up to 14 days if stored at 4°C.

Medium: Antibiotic medium 2. Adjust final pH before inoculation to pH 7.5 (pH lower than medium 11).

Diluent: Phosphate buffer, pH 8.0 (solution 3)

Standard: As in method 1 (Section 9.20.1)

Dose levels: 20 IU/ml and 5 IU/ml (dose ratio 4:1)

Proceed as in method 1 (Section 9.20.1).

Incubation: 37°C overnight

Remarks: This method is suitable for the assay of neomycin in the presence of streptomycin.

9.21 Novobiocin

Organism: *Bacillus subtilis* ATCC 12432, NCIB 8993

Prepare a concentrated spore suspension as described in Section 2.8.

Inoculum: Mix about 1.0% (v/v) of the calibrated spore suspension to the assay medium held at 60°C.

Medium: Antibiotic medium 1. Adjust final pH before inoculation to pH 6.5.

Diluents: (1) Phosphate buffer, pH 7.5 (solution 20). (2) Phosphate buffer, pH 6.0 (solution 1).

Standard: A working standard of sodium novobiocin of known chemical purity. Express results in terms of the anhydrous compound (e.g., 950 μg/mg).

Dose levels: 10 μg/ml and 5 μg/ml (dose ratio 2:1)

After drying the working standard for 3 hours at 60°C in a vacuum oven with a residual pressure of 5 mm Hg, accurately weigh about 110 mg and quantitatively dissolve in 100 ml of phosphate buffer, pH 7.5. Prepare further dilutions in pH 6.0 phosphate buffer. For example:

$$110 \text{ mg} \rightarrow 100 \text{ ml} : 10 \text{ ml} \rightarrow 100 \text{ ml} : 10 \text{ ml} \rightarrow 100 \text{ ml}$$
$$: 10 \text{ ml} \rightarrow 200 \text{ ml}$$

Store the primary stock solution of the working standard at 4°C for not longer than 7 days.

Incubation: 30°C overnight

9.22 Nystatin

Organism: *Saccharomyces cerevisiae* NCYC 81

Maintain the test organism by regular subculturing on Sabouraud's agar slants and incubating at 30°C for 48 hours.

Inoculum: Suspend the growth from a freshly grown slant in about 50 ml of sterile saline and mix about 1.0% (v/v) of the calibrated suspension to the assay medium held at just below 50°C. The inoculum is usable for about 3 weeks if stored at 4°C.

Medium: Antibiotic medium 41. Adjust final pH before inoculation to pH 6.5. Before adjusting pH, add a solution of penicillin G to give a concentration of 2 U/ml, which will prevent bacterial contamination of the medium.

Diluents: (1) Dimethylformamide (DMF). (2) Phosphate buffer, pH 6.0 (solution 1).

Standard: A working standard of known chemical purity, calibrated directly or indirectly against the international standard. Store the working standard at -20°C or below.

Dose levels: 80 IU/ml and 40 IU/ml (dose ratio 2:1)

For example, accurately weigh about 100 mg of the working standard, having a potency of 4390 IU/mg, and quantitatively dissolve in 100 ml of DMF. Dilute 20 ml to 50 ml with DMF and further dilute to plating-out levels with pH 6.0 phosphate buffer, keeping the DMF concentration constant:

$$100 \text{ mg} \rightarrow 100 \text{ ml (DMF)}: 20 \text{ ml} \rightarrow 50 \text{ ml (DMF)}: 10 \text{ ml} \rightarrow 200 \text{ ml (solution 1)}$$
$$: 5 \text{ ml} + 5 \text{ ml (DMF)} \rightarrow$$
$$200 \text{ ml (solution 1)}$$

Incubation: 37°C overnight

Remarks: Protect solutions from light and do not store, but prepare fresh as required.

9.23 Oxytetracycline

Organism: *Bacillus subtilis* ATCC 6633, NCIB 8533

Prepare a concentrated spore suspension as detailed in Section 2.8.

Inoculum: Mix about 1.0% (v/v) of the calibrated spore suspension to the assay medium held at 60°C.

Medium: Antibiotic medium 40. Final pH before inoculation adjusted to pH 6.7.

Diluents: (1) 2% (v/v) HCl in methanol (solution 22). (2) Phosphate buffer, pH 4.5 (solution 4).

Standard: A working standard of oxytetracycline hydrochloride of known chemical purity, calibrated directly or indirectly against the international standard.

Dose levels: 2.0 IU/ml and 1.0 IU/ml (dose ratio 2:1)

For example, accurately weigh about 110 mg of the working standard, having a potency of 892 IU/mg, and quantitatively dissolve in 100 ml of solution 22. Further dilute to plating-out levels with phosphate buffer, pH 4.5 (solution 4):

$$110 \text{ mg} \rightarrow 100 \text{ ml}: 5 \text{ ml} \rightarrow 100 \text{ ml}: 10 \text{ ml} \rightarrow 50 \text{ ml}: 10 \text{ ml} \rightarrow 50 \text{ ml}$$
$$: 10 \text{ ml} \rightarrow 100 \text{ ml}$$

Store the primary stock solution of the working standard for not longer than 7 days at 4°C.

Incubation: 30°C overnight

9.24 Penicillin G

9.24.1 Method 1

Organism: *Bacillus subtilis* ATCC 6633, NCIB 8533

Prepare a concentrated spore suspension as detailed in Section 2.8.

Inoculum: Mix about 1.0% (v/v) of the calibrated spore suspension to the assay medium held at 60°C.

Medium: Antibiotic medium 2. Final pH before inoculation adjusted to pH 6.5.

Diluents: (1) Phosphate buffer, pH 6.0 (solution 1). (2) Phosphate buffer, pH 7.5 (solution 20).

Standard: A working standard of sodium benzyl penicillin of known chemical purity and potency

Dose levels: 0.5 IU/ml and 0.25 IU/ml (dose ratio 2:1)

For example, accurately weigh about 120 mg of the working standard, having a potency of 1655 U/mg after drying at 100°C for 3 hours. Dissolve quantitatively in 100 ml pH 6.0 phosphate buffer, then further dilute in pH 7.5 phosphate buffer to plating-out levels. Typically:

$$120 \text{ mg} \rightarrow 100 \text{ ml}:5 \text{ ml} \rightarrow 500 \text{ ml}:5 \text{ ml} \rightarrow 200 \text{ ml}:50 \text{ ml} \rightarrow 100 \text{ ml}$$

Store the primary stock solution of the working standard at 4°C for not longer than 4 days.

Incubation: 30°C overnight

Remarks: This method is suitable for the assay of benzyl penicillin and procaine penicillin. Consult the USP or the BP for the solubilization of procaine penicillin.

9.24.2   Method 2

Organism: Bacillus subtilis, not deposited with ATCC or NCIB; Glaxo Group Research Culture Collection No. 1768E

Prepare a concentrated spore suspension as detailed in Section 2.8.

Inoculum: Mix about 1.0% (v/v) of the calibrated spore suspension to the assay medium held at 60°C.

Medium: Antibiotic medium 42. Final pH before inoculation is adjusted to pH 7.5.

Diluents: (1) Phosphate buffer, pH 6.0 (solution 1). (2) Phosphate buffer, pH 7.5 (solution 20).

Standard: As in method 1 (Section 9.24.1)

Dose levels: 0.04 U/ml and 0.02 U/ml (dose ratio 2:1)

Proceed as in method 1 (Section 9.24.1), but dilute to plating-out levels typically as follows:

$$120 \text{ mg} \rightarrow 100 \text{ ml}:5 \text{ ml} \rightarrow 500 \text{ ml}:10 \text{ ml} \rightarrow 250 \text{ ml}:5 \text{ ml} \rightarrow 100 \text{ ml and}$$
$$:5 \text{ ml} \rightarrow 200 \text{ ml}$$

Store the primary stock solution of the working standard at 4°C for not longer than 4 days.

Incubation: 37°C overnight

Remarks: This method is suitable for the determination of trace levels of penicillin, as the method is capable of detecting penicillin levels of 0.01 U/ml. Use penicillin-free glassware. (Wash in weak penicillinase solution, followed by a thorough rinsing with penicillin-free distilled water.)

9.24.3   Method 3

Organism: Bacillus subtilis, not deposited with ATCC or NCIB; Glaxo Group Research Culture Collection No. MB 32 SDR

Prepare a concentrated spore suspension as detailed in Section 2.8.

Inoculum: Mix about 1.0% (v/v) of the calibrated spore suspension to the assay medium held at 60°C.

Medium: Antibiotic medium 43. Add about 0.1 ml of a sterile 40% (w/v) glucose solution to each 100 ml of assay medium, and adjust to pH 7.0 before inoculation.

Diluents: (1) Phosphate buffer, pH 6.0 (solution 1). (2) Phosphate buffer, pH 7.5 (solution 20).

Standard: As in method 1 (Section 9.24.1)

Dose levels: 1.0 U/ml and 0.5 U/ml (dose ratio 2 : 1)

Proceed as in method 1 (Section 9.24.1), but dilute to plating-out levels using solution 20, typically as follows:

$$120 \text{ mg} \rightarrow 100 \text{ ml} : 5 \text{ ml} \rightarrow 500 \text{ ml} : 5 \text{ ml} \rightarrow 100 \text{ ml and}$$
$$: 5 \text{ ml} \rightarrow 200 \text{ ml}$$

Store the primary stock solution of the working standard at 4°C for not longer than 4 days.

Incubation: 30°C overnight

Remarks: This method is suitable for the determination of penicillin in the presence of streptomycin or dihydrostreptomycin.

9.25   Penicillin V

Organism: *Bacillus subtilis* ATCC 6633, NCIB 8533

Prepare a concentrated spore suspension as detailed in Section 2.8.

Inoculum: Mix about 1.0% (v/v) of the calibrated spore suspension to the assay medium held at 60°C.

Medium: Antibiotic medium 2. Final pH before inoculation is adjusted to pH 6.5.

Diluents: (1) Acetone AR. (2) Phosphate buffer, pH 7.5 (solution 20).

Standard: A working standard of known chemical purity and potency which was preferably standardized against the USP Reference Standard of penicillin V.

Dose levels: 0.5 U/ml and 0.25 U/ml (dose ratio 2 : 1)

Typically, weigh accurately about 120 mg of the working standard having a potency of 1538 U/mg, dissolve by adding 10 ml of acetone and 20 ml of solution 20, then dilute to 1000 ml with solution 20 and further dilute to plating-out levels using solution 20. For example:

$$120 \text{ mg} \rightarrow 1000 \text{ ml} : 5 \text{ ml} \rightarrow 150 \text{ ml} : 10 \text{ ml} \rightarrow 100 \text{ ml and}$$
$$10 \text{ ml} \rightarrow 200 \text{ ml}$$

Store the primary stock solution of the working standard at 4°C for not longer than 7 days.

Incubation: 30°C overnight

9.26   Phenylmercuric Acetate

Organism: *Micrococcus luteus* ATCC 10240b, NCIB 8994

Maintain the test organism by regular subculturing on slants of the assay medium and incubating at 30°C for 48 hours.

Inoculum: Suspend the growth from a freshly grown slant in 10 ml of sterile saline and mix about 1.0% (v/v) of the calibrated suspension to the assay medium held at 50°C. The inoculum is usable for up to 14 days if stored at 4°C.

Medium: Antibiotic medium 42. Final pH before inoculation is adjusted to pH 7.9.

Diluent: Distilled water

Standard: A working standard of known chemical purity. Typically, weigh about 100 mg of the working standard and quantitatively dissolve in about 40 ml of warm distilled water. Allow to cool and make up volume to 1000 ml, then dilute further to plating-out levels using distilled water. For example:

$$100 \text{ mg} \rightarrow 1000 \text{ ml} : 5 \text{ ml} \rightarrow 100 \text{ ml and}$$
$$5 \text{ ml} \rightarrow 200 \text{ ml}$$

Dose levels: 0.0005% (w/v) and 0.00025% (w/v) (dose ratio 2 : 1)
Incubation: 37°C overnight
Remarks:
1. Phenylmercuric nitrate can be assayed by the same method.
2. Protect solutions from light.
3. The primary stock solution of the working standard can be stored at 4°C for up to 1 week.
4. For further details refer to Vincent and Shaw (1972).

9.27 Streptomycin

Organism: *Bacillus subtilis* NCIB 8533
Prepare a concentrated spore suspension as detailed in Section 2.8.
Inoculum: Mix about 1.0% (v/v) of the calibrated spore suspension to the assay medium held at 60°C.
Medium: Antibiotic medium 40. Final pH before inoculation is adjusted to pH 7.5.
Diluent: Phosphate buffer, pH 7.5 (solution 20)
Standard: A working standard of known chemical composition and purity of streptomycin sulfate, calibrated directly or indirectly against the international standard.
Dose levels: 2.0 IU/ml and 1.0 IU/ml (dose ratio 2 : 1)
Typically, weigh about 125 mg of the working standard and dry for 3 hours at 60°C and a pressure of 5 mm Hg or less. Quantitatively dissolve in 1000 ml of solution 20 and further dilute to plating-out levels using solution 20. For example:

$$125 \text{ mg} \rightarrow 1000 \text{ ml} : 10 \text{ ml} \rightarrow 100 \text{ ml} : 10 \text{ ml} \rightarrow 50 \text{ ml and}$$
$$: 10 \text{ ml} \rightarrow 100 \text{ ml}$$

Store the primary stock solution at 4°C for not longer than 4 weeks.
Incubation: 37°C overnight
Remarks: The same method is also suitable for the assay of streptomycin in the presence of neomycin if the plates are incubated at 30°C.

9.28 Tetracycline

Organism: *Bacillus subtilis* NCIB 8533
Prepare a concentrated spore suspension as detailed in Section 2.8.
Inoculum: Mix about 1.0% (v/v) of the calibrated spore suspension to the assay medium held at 60°C.
Medium: Antibiotic medium 2. Final pH before inoculation is adjusted to pH 6.5. Before inoculation add about 0.5% (v/v) of a sterile 40% (w/v) solution of glucose to the assay medium.
Diluents: (1) 0.01 *N* HCl. (2) Phosphate buffer, pH 4.5 (solution 4).
Standard: A working standard of tetracycline of known chemical composition and purity, calibrated directly or indirectly against the international standard.

Dose levels: 1.0 IU/ml and 0.5 IU/ml (dose ratio 2 : 1)

Typically, weigh about 50 mg of the working standard and dissolve quantitatively in 50 ml of 0.01 *N* HCl. Further dilute to plating-out levels using solution 4. For example:

$$50 \text{ mg} \rightarrow 50 \text{ ml} : 5 \text{ ml} \rightarrow 500 \text{ ml} : 10 \text{ ml} \rightarrow 100 \text{ ml and}$$
$$: 10 \text{ ml} \rightarrow 200 \text{ ml}$$

Store the primary stock solution of the working standard at 4°C for not longer than 7 days.

Incubation: 30°C overnight

9.29   Thiomersal

Organism: *Micrococcus luteus* ATCC 10240b, NCIB 8994

Maintain the test organism by regular subculturing on slants of the assay medium and incubating at 30°C for 48 hours.

Inoculum: Suspend the growth from a freshly grown slant in 10 ml of sterile saline and mix about 1.0% (v/v) of the calibrated suspension to the assay medium held at 50°C. The inoculum is usable for up to 14 days if stored at 4°C.

Medium: Antibiotic medium 42. Final pH before inoculation is adjusted to pH 7.9.

Diluent: Distilled water

Standard: A working standard of known chemical purity. Typically, weigh about 100 mg of the working standard and quantitatively dissolve in 1000 ml of distilled water. Further dilute with distilled water to plating-out levels. For example:

$$100 \text{ mg} \rightarrow 1000 \text{ ml} : 5 \text{ ml} \rightarrow 200 \text{ ml} : 50 \text{ ml} \rightarrow 100 \text{ ml}$$

The primary stock solution of the working standard can be used for up to 7 days if stored at 4°C.

Dose levels: 0.00025 μg/ml and 0.000125 μg/ml (dose ratio 2 : 1)

Remarks: For further details refer to Vincent and Shaw (1972).

9.30   Tobramycin

Organism: *Bacillus pumilus* ATCC 14884, NCTC 8241

Prepare a concentrated spore suspension as detailed in Section 2.8.

Inoculum: Mix about 1.0% (v/v) of the calibrated spore suspension to the assay medium held at 60°C.

Medium: Antibiotic medium 5. Final pH before inoculation is adjusted to pH 7.9 if necessary.

Diluents: (1) Distilled water. (2) Phosphate buffer, pH 8.0 (solution 3).

Standard: A working standard of known chemical composition and purity, calibrated directly or indirectly against the international standard.

Dose levels: 0.5 IU/ml and 0.25 IU/ml (dose ratio 2 : 1)

Typically, weigh about 110 mg of the working standard and dissolve it quantitatively in 100 ml of distilled water. Further dilute to plating-out levels using solution 3, for example:

$$110 \text{ mg} \rightarrow 100 \text{ ml} : 5 \text{ ml} \rightarrow 500 \text{ ml} : 10 \text{ ml} \rightarrow 100 \text{ ml} : 10 \text{ ml} \rightarrow 100 \text{ ml and}$$
$$: 10 \text{ ml} \rightarrow 200 \text{ ml}$$

Store the stock solution of the working standard at 4°C for not longer than 4 days.

Incubation: 37°C overnight

Remarks: This method is suitable for the assay of tobramycin even in the presence of cefuroxime. Ratio of tobramycin to cefuroxime = 1:5.5.

9B   CODE OF FEDERAL REGULATIONS (21 CFR) METHODS

9.31   Amoxicillin

Organism: *Micrococcus luteus* ATCC 9341
Maintain on slants of medium 1 and incubate for 24 hours at 32° – 35°C. Wash the growth from the agar slant with 3 ml of sterile USP saline TS onto a large agar surface such as a Roux bottle containing 250 ml of medium 1. Spread the suspension over the entire agar surface using sterile glass beads, if necessary. Incubate the Roux bottle at 32° – 35°C for 24 hours and wash the resulting growth from the agar surface with 50 ml of sterile USP saline TS. Dilute the suspension 1:20 and store refrigerated for not longer than 1 week.
Inoculum: 0.5 ml per 100 ml of medium 11
Medium: Base layer, 21 ml of medium 11 (uninoculated). Seed layer, 4 ml of medium 11.
Standard: *Not dried.* Prepare a 1 mg/ml stock solution of the working standard in distilled water and store refrigerated for not longer than 7 days. Further dilute with solution 3.
Dose levels: 0.064, 0.080, 0.100, 0.125, and 0.156 μg/ml
Prepare the standard response line simultaneously with the sample solution.
Incubation: 32° – 35°C

9.32   Amphomycin

Organism: *Micrococcus luteus* ATCC 14452 (resistant to neomycin)
Maintain on slants of medium 1 containing about 100 μg/ml of neomycin; otherwise proceed as described in Section 9.31, but incubate the Roux bottles for 48 hours. Dilute 1:35 and store refrigerated for not longer than 4 weeks.
Inoculum: 0.5 ml per 100 ml of medium 1
Medium: Base layer, 21 ml of medium 2. Seed layer, 4 ml of medium 1.
Standard: Transfer about 100 mg of the working standard to a tared weighing bottle equipped with a ground-glass stopper. Weigh the bottle and place it in a vacuum oven, tilting the stopper on its side so that there is no closure during the drying period. Dry at a temperature of 60°C and a pressure of 5 mm Hg or less for 3 hours. At the end of the drying period, fill the vacuum oven with air dried by passing it through a drying agent such as sulfuric acid or silica gel. Replace the stopper and place the weighing bottle in a desiccator over a desiccating agent, such as phosphorus pentoxide or silica gel, allow to cool to room temperature, and reweigh. Calculate the percentage loss. Prepare a 0.1 mg/ml stock solution in Solution 3, allow to stand overnight at room temperature to ensure complete solubilization, and store refrigerated for not longer than 14 days. Further dilute with Solution 3.
Dose levels: 6.4, 8.0, 10.0, 12.5, and 15.6 μg/ml
Incubation: 36° – 37.5°C

9.33   Amphotericin B

Organism: *Saccharomyces cerevisiae* ATCC 9763
Proceed as described in Section 9.31, except incubate the slants of medium 19 at 30°C for 24

hours and incubate the Roux bottles containing medium 19 at 30°C for 48 hours. Dilute 1 : 30 and store refrigerated for not longer than 4 weeks.

Inoculum: 1.0 ml per 100 ml of medium 19

Medium: No base layer. Seed layer, 8 ml of medium 19.

Standard: Dry the working standard as described in Section 9.32. Prepare a 1 mg/ml stock solution in dimethyl sulfoxide (DMS) and use the same day. Further dilute with DMS to give concentrations of 12.8, 16.0, 20.0, 25.0, and 31.2 μg/ml, and finally, dilute each of these solutions 1 : 20 (5 ml → 100 ml) with solution 10.

Dose levels: 0.64, 0.80, 1.00, 1.25, and 1.56 μg/ml

Prepare the standard response line simultaneously with the sample solution.

Incubation: 29° – 31°C

9.34   Ampicillin

Organism: Micrococcus luteus ATCC 9341

For culture maintenance and suspension preparation proceed as described in Section 9.31. Dilute 1 : 40 and store refrigerated for not longer than 2 weeks.

Inoculum: 0.5 ml per 100 ml of medium 11

Medium: Base layer, 21 ml of medium 11 (uninoculated). Seed layer, 4 ml of medium 11.

Standard: Not dried. Prepare a 0.1 mg/ml stock solution in distilled water and store refrigerated for not longer than 1 week. Further dilute with solution 3.

Dose levels: 0.064, 0.080, 0.100, 0.125, and 0.156 μg/ml

Prepare the standard response line simultaneously with the sample solution.

Incubation: 32° – 35°C

9.35   Bacitracin

9.35.1   Method 1

Organism: Micrococcus luteus ATCC 7468

Maintain and prepare a suspension as described in Section 9.31. Dilute 1 : 30 and store refrigerated for not longer than 2 weeks.

Inoculum: 0.3 ml per 100 ml of medium 1

Medium: Base layer, 21 ml of medium 2. Seed layer, 4 ml of medium 1.

Standard: Dry the zinc bacitracin working standard as described in Section 9.32. Prepare a 100 IU/ml stock solution in 0.01 N hydrochloric acid (HCl) and use the same day. Further dilute with solution 1.

Dose levels: 0.64, 0.80, 1.0, 1.25, and 1.56 IU/ml

Incubation: 32° – 35°C

9.35.2   Method 2

Organism: Micrococcus luteus ATCC 10240

Maintain and prepare a suspension as described in Section 9.31. Dilute 1 : 35 and store refrigerated for not longer than 4 weeks.

Inoculum: 0.3 ml per 100 ml of medium 1

Medium, standard, dose levels, and *incubation* are as in Method 1 (Section 9.35.1).

9.36   Bleomycin

Organism: *Mycobacterium smegmatis* ATCC 607
Maintain by weekly subculturing on slants of medium 36 and incubate at 37°C for 48 hours. Wash off the growth from the agar slant into a 500-ml Erlenmeyer flask containing 100 ml of medium 34 and 50 g of glass beads. Agitate the culture by rotation at a speed of 130 cycles per minute and a radius of 3.5 cm at 27°C for 5 days. Determine the amount of suspension needed for 100 ml medium 35 (~ 1 ml). Store the suspension refrigerated for not longer than 2 weeks.
Medium: Base layer, 10 ml of uninoculated medium 35. Seed layer, 6 ml of medium 35.
Standard: Proceed as described in Section 9.32, except dry at 25°C for 4 hours. Prepare a 2 U/ml stock solution in solution 16 and store refrigerated for not longer than 2 weeks. Further dilute with solution 16.
Dose levels: 0.01, 0.02, 0.04, 0.08, and 0.16 U/ml
Incubation: 32°–35°C
Remarks: Suspend the test organism in medium 34 instead of saline for the determination of light transmittance.

9.37   Carbenicillin

Organism: *Pseudomonas aeruginosa* ATCC 25619
Maintain and prepare a suspension as described in Section 9.31, except incubate at 36°–37.5°C. Dilute 1 : 25 and store refrigerated for not longer than 2 weeks.
Inoculum: 0.5 ml per 100 ml of medium 10
Medium: Base layer, 21 ml of medium 9. Seed layer, 4 ml of medium 10.
Standard: Not dried. Prepare a 1 mg/ml stock solution in solution 1 and store refrigerated for not longer than 2 weeks. Further dilute in solution 1.
Dose levels: 12.8, 16.0, 20.0, 25.0, and 31.2 μg/ml
Incubation: 36°–37.5°C

9.38   Cefactor

Organism: *Staphylococcus aureus* ATCC 6538P
Maintain and prepare a suspension as described in Section 9.31. Dilute 1 : 20 and store refrigerated for not longer than 1 week.
Inoculum: 0.05 ml per 100 ml of medium 1
Medium: Base layer, 21 ml of medium 2. Seed layer, 5 ml of medium 1.
Standard: Not dried. Prepare a stock solution of 1 mg/ml in solution 1 and store refrigerated for not longer than 1 day. Further dilute with solution 1.
Dose levels: 3.2, 4.0, 5.0, 6.25, and 7.81 μg/ml
Incubation: 36°–37.5°C

9.39   Cefadroxil

Organism: *Staphylococcus aureus* ATCC 6538P
Maintain and prepare a suspension as described in Section 9.31. Dilute 1 : 20 and store refrigerated for not longer than 1 week.

Inoculum: 0.05 ml per 100 ml of medium 1
Medium: Base layer, 21 ml of medium 2. Seed layer, 4 ml of medium 1.
Standard: Not dried. Prepare a 1 mg/ml stock solution in solution 1. *Use the same day.*
Further dilute in solution 1.
Dose levels: 12.8, 16.0, 20.0, 25.0, and 31.2 μg/ml
Incubation: 36° – 37.5°C

9.40 Cefamandole

Organism: Staphylococcus aureus ATCC 6538P
Maintain and prepare a suspension as described in Section 9.31. Dilute 1:20 and store
refrigerated for not longer than 1 week.
Inoculum: 0.06 ml per 100 ml of medium 1
Medium: Base layer, 21 ml of medium 2. Seed layer, 5 ml of medium 1.
Standard: Not dried. Prepare a 1 mg/ml stock solution in solution 3. Allow to hydrolyze in
a 37°C water bath for 1 hour. Store refrigerated for not longer than 1 day. Further dilute with
solution 1.
Dose levels: 1.28, 1.60, 2.00, 2.50, and 3.12 μg/ml
Incubation: 36° – 37.5°C

9.41 Cefazolin

Organism: Staphylococcus aureus ATCC 6538P
Maintain and prepare a suspension as described in Section 9.31. Dilute 1:20 and store
refrigerated for not longer than 1 week.
Inoculum: 0.05 ml per 100 ml of medium 1
Medium: Base layer, 21 ml of medium 2. Seed layer, 4 ml of medium 1.
Standard: Not dried. Prepare a 10,000 μg/ml stock solution in solution 6 and store
refrigerated for not longer than 5 days. Further dilute with solution 1.
Dose levels: 0.64, 0.80, 1.00, 1.25, and 1.56 μg/ml
Incubation: 32° – 35°C

9.42 Cefotaxime

Organism: Staphylococcus aureus ATCC 6538P
See Section 9.31 for maintenance and preparation of suspension. Dilute 1:20 and store
refrigerated for not longer than 1 week.
Inoculum: 0.1 ml per 100 ml of medium 1
Medium: Base layer, 21 ml of medium 2. Seed layer, 5 ml of medium 1.
Standard: Not dried. Prepare a 1 mg/ml stock solution in solution 1 and *use the same day.*
Further dilute with solution 1.
Dose levels: 6.4, 8.0, 10.0, 12.5, and 15.6 μg/ml
Incubation: 36° – 37.5°C

9.43 Cefoxitin

Organism: *Staphylococcus aureus* ATCC 6538P
For maintenance and preparation of suspension see Section 9.31. Dilute 1:20 and store refrigerated for not longer than 1 week.
Inoculum: 0.1 ml per 100 ml of medium 1.
Medium: Base layer, 21 ml of medium 2. Seed layer, 5 ml of medium 1.
Standard: *Not dried.* Prepare a 1 mg/ml stock solution in solution 1 and *use the same day.* Further dilute with solution 1.
Dose levels: 12.8, 16.0, 20.0, 25.0, and 31.2 μg/ml
Incubation: 36°–37.5°C

9.44 Cephalexin

Organism: *Staphylococcus aureus* ATCC 6538P
For maintenance and suspension preparation see Section 9.31. Dilute 1:20 and store refrigerated for not longer than 1 week.
Inoculum: 0.05 ml per 100 ml of medium 1
Medium: Base layer, 21 ml of medium 2. Seed layer, 4 ml of medium 1.
Standard: *Not dried.* Prepare a 1 mg/ml stock solution in solution 1 and store refrigerated for not longer than 7 days. Further dilute with solution 1.
Dose levels: 12.8, 16.0, 20.0, 25.0, and 31.2 μg/ml
Incubation: 32°–35°C

9.45 Cephaloglycin

Organism: *Staphylococcus aureus* ATCC 6538P
For maintenance and suspension preparation see Section 9.31. Dilute 1:20 and store refrigerated for not longer than 1 week.
Inoculum: 0.2 ml per 100 ml of medium 1
Medium: Base layer, 21 ml of medium 2. Seed layer, 4 ml of medium 1.
Standard: *Not dried.* Prepare a 100 μg/ml stock solution in distilled water and store refrigerated for not longer than 1 week. Further dilute with solution 4.
Dose levels: 6.4, 8.0, 10.0, 12.5, and 15.6 μg/ml
Incubation: 32°–35°C

9.46 Cephaloridine

Organism: *Staphylococcus aureus* ATCC 6538P
For maintenance and suspension preparation see Section 9.31. Dilute 1:20 and store refrigerated for not longer than 1 week.
Inoculum: 0.1 ml per 100 ml of medium 1
Medium: Base layer, 21 ml of medium 2. Seed layer, 4 ml of medium 1.

Standard: Proceed as described in Section 9.32. Prepare a 1 mg/ml stock solution in solution 1 and store refrigerated for not longer than 5 days. Further dilute with solution 1.

Dose levels: 0.64, 0.80, 1.00, 1.25, and 1.56 μg/ml

Incubation: 32°–35°C

9.47   Cephalothin

Organism: *Staphylococcus aureus* ATCC 6538P

Maintain and prepare a suspension as described in Section 9.31. Dilute 1:20 and store refrigerated for not longer than 1 week.

Inoculum: 0.1 ml per 100 ml of medium 1

Medium: Base layer, 21 ml of medium 2. Seed layer, 4 ml of medium 1.

Standard: Proceed as described in Section 9.32. Prepare a 1 mg/ml stock solution in solution 1 and store refrigerated for not longer than 5 days. Further dilute with solution 1.

Dose levels: 0.64, 0.80, 1.00, 1.25, and 1.56 μg/ml

Incubation: 32°–35°C

9.48   Cephapirin

Organism: *Staphylococcus aureus* ATCC 6538P

Maintain and prepare a suspension as described in Section 9.31. Dilute 1:20 and store refrigerated for not longer than 1 week.

Inoculum: 0.08 ml per 100 ml of medium 1

Medium: Base layer, 21 ml of medium 2. Seed layer, 4 ml of medium 1.

Standard: *Not dried.* Prepare a 1 mg/ml stock solution in solution 1 and store refrigerated for not longer than 3 days. Further dilute with solution 1.

Dose levels: 0.64, 0.80, 1.00, 1.25, and 1.56 μg/ml

Incubation: 32°–35°C

9.49   Cephradine

Organism: *Staphylococcus aureus* ATCC 6538P

Maintain and prepare a suspension as described in Section 9.31. Dilute 1:20 and store refrigerated for not longer than 1 week.

Inoculum: 0.05 ml per 100 ml of medium 1

Medium: Base layer, 21 ml of medium 2. Seed layer, 4 ml of medium 1.

Standard: *Not dried.* Prepare a 1 mg/ml stock solution in solution 1 and store refrigerated for not longer than 5 days. Further dilute with solution 1.

Dose levels: 6.4, 8.0, 10.0, 12.5, and 15.6 μg/ml

Incubation: 32°–35°C

9.50   Clindamycin

Organism: *Micrococcus luteus* ATCC 9341

Maintain and prepare a suspension as described in Section 9.31. Dilute 1:40 and store refrigerated for not longer than 2 weeks.

Inoculum: 1.5 ml per 100 ml of medium 11
Medium: Base layer, 21 ml of uninoculated medium 11. Seed layer, 4 ml of medium 11.
Standard: Not dried. Prepare a 1 mg/ml stock solution in distilled water and store refrigerated for not longer than 1 month. Further dilute with solution 3.
Dose levels: 0.64, 0.80, 1.00, 1.25, and 1.56 µg/ml
Incubation: 36° – 37.5°C

9.51 Cloxacillin

Organism: Staphylococcus aureus ATCC 6538P
Maintain and prepare a suspension as described in Section 9.31. Dilute 1:20 and store refrigerated for not longer than 1 week.
Inoculum: 0.1 ml per 100 ml of medium 1
Medium: Base layer, 21 ml of medium 2. Seed layer, 4 ml of medium 1.
Standard: Not dried. Prepare a 1 mg/ml stock solution in solution 1 and store refrigerated for not longer than 7 days. Further dilute with solution 1.
Dose levels: 3.20, 4.00, 5.00, 6.25, and 7.81 µg/ml
Incubation: 32° – 35°C

9.52 Colistimethate Sodium

Organism: Bordetella bronchiseptica ATCC 4617
Maintain and prepare a suspension as described in Section 9.31. Dilute 1:20 and store refrigerated for not longer than 2 weeks.
Inoculum: 0.1 ml per 100 ml of medium 10
Medium: Base layer, 21 ml of medium 9. Seed layer, 4 ml of medium 10.
Standard: Proceed as described in Section 9.32. Prepare a 10,000 µg/ml stock solution in distilled water and *use the same day.* Further dilute with solution 6.
Dose levels: 0.64, 0.80, 1.00, 1.25, and 1.56 µg/ml.
Incubation: 36° – 37.5°C

9.53 Colistin

Organism: Bordetella bronchiseptica ATCC 4617
Maintain and prepare a suspension as described in Section 9.31. Dilute 1:20 and store refrigerated for not longer than 2 weeks.
Inoculum: 0.1 ml per 100 ml of medium 10
Medium: Base layer, 21 ml of medium 9. Seed layer, 4 ml of medium 10.
Standard: Proceed as described in Section 9.32. Prepare a 10,000 µg/ml stock solution in distilled water and store refrigerated for not longer than 2 weeks. Further dilute with solution 6.
Dose levels: 0.64, 0.80, 1.00, 1.25, and 1.56 µg/ml
Incubation: 36° – 37.5°C

9.54 Cyclacillin

Organism: Micrococcus luteus ATCC 9341

Maintain and prepare a suspension as described in Section 9.31. Dilute 1:40 and store refrigerated for not longer than 2 weeks.

Inoculum: 0.5 ml per 100 ml of medium 11

Medium: Base layer 21 ml of uninoculated medium 11. Seed layer, 4 ml of medium 11.

Standard: *Not dried.* Prepare a 1 mg/ml stock solution in distilled water and store refrigerated for not longer than 1 day. Further dilute with solution 3.

Dose levels: 0.64, 0.80, 1.00, 1.25, and 1.56 μg/ml

Incubation: 36°–37.5°C

Remarks: Prepare the standard response line simultaneously with the sample solution.

9.55   Dactinomycin

Organism: *Bacillus subtilis* ATCC 6633

Grow on slants of medium 1 at 32°–35°C for 24 hours and wash off the growth with 3 ml of sterile USP saline TS. Spread the suspension over a large agar surface, such as in a Roux bottle containing 250 ml of medium 32, and incubate at 32°–35°C for at least 5 days. Wash the resulting growth with 50 ml of sterile USP saline TS, centrifuge, and decant the supernatant liquid. Resuspend the sediment with 50–70 ml of sterile USP saline TS and heat the suspension for 30 minutes at 70°C. Store the standardized suspension under refrigeration and restandardize every 6 months, if necessary.

Inoculum: Use test plates to determine the amount of inoculum needed for each 100 ml of medium 5.

Medium: Base layer, 10 ml of uninoculated medium 5. Seed layer, 4 ml of medium 5.

Standard: Proceed as described in Section 9.32. Prepare a 10,000 μg/ml stock solution in methanol and store refrigerated for not longer than 3 months. Further dilute with solution 3.

Dose levels: 0.50, 0.71, 1.00, 1.41, and 2.00 μg/ml

Incubation: 36°–37.5°C

9.56   Dicloxacillin

Organism: *Staphylococcus aureus* ATCC 6538P

Maintain and prepare a suspension as described in Section 9.31. Dilute 1:20 and store refrigerated for not longer than 1 week.

Inoculum: 0.1 ml per 100 ml of medium 1

Medium: Base layer, 21 ml of medium 2. Seed layer, 4 ml of medium 1.

Standard: *Not dried.* Prepare a 1 mg/ml stock solution in solution 1 and store refrigerated for not longer than 7 days. Further dilute with solution 1.

Dose levels: 3.2, 4.00, 5.00, 6.25, and 7.81 μg/ml

Incubation: 32°–35°C

9.57   Dihydrostreptomycin

Organism: *Bacillus subtilis* ATCC 6633

Prepare, standardize, and store the spore suspension as described in Section 9.55.

Inoculum: Use test plates to determine the amount of inoculum needed for each 100 ml of medium 5.

Medium: Base layer, 21 ml of uninoculated medium 5. Seed layer, 4 ml of medium 5.

Standard: Proceed as described in Section 9.32, except dry at 100°C for 4 hours. Prepare a 1 mg/ml stock solution in solution 3 and store refrigerated for not longer than 30 days. Further dilute with solution 3.

Dose levels: 0.64, 0.80, 1.00, 1.25, and 1.56 μg/ml

Incubation: 36° – 37.5°C

9.58   Erythromycin

Organism: *Micrococcus luteus* ATCC 9341

Maintain and prepare a suspension as described in Section 9.31. Dilute 1 : 40 and store refrigerated for not longer than 2 weeks.

Inoculum: 1.5 ml per 100 ml of medium 11

Medium: Base layer, 21 ml of uninoculated medium 11. Seed layer, 4 ml of medium 11.

Standard: Dry as described in Section 9.31. Prepare a 10,000 μg/ml stock solution in methanol and store refrigerated for not longer than 14 days. Further dilute with solution 3.

Dose levels: 0.64, 0.80, 1.00, 1.25, and 1.56 μg/ml

Incubation: 32° – 35°C

9.59   Gentamicin

Organism: *Staphylococcus epidermidis* ATCC 12228

Maintain and prepare a suspension as described in Section 9.31. Dilute 1 : 14 and store refrigerated for not longer than 1 week.

Inoculum: 0.03 ml per 100 ml of medium 11

Medium: Base layer, 21 ml of uninoculated medium 11. Seed layer, 4 ml of medium 11.

Standard: Dry as described in Section 9.32, except use a temperature of 110°C. Prepare a 1 mg/ml stock solution in solution 3 and store refrigerated for not longer than 1 month. Further dilute with solution 3.

Dose levels: 0.064, 0.080, 0.100, 0.125, and 0.156 μg/ml

Incubation: 36° – 37.5°C

9.60   Kanamycin B

Organism: *Bacillus subtilis* ATCC 6633

Prepare, standardize, and store the spore suspension as described in Section 9.55.

Inoculum: Use test plates to determine the amount of inoculum needed for each 100 ml of medium 5.

Medium: Base layer, 21 ml of uninoculated medium 5. Seed layer, 4 ml of medium 5.

Standard: *Not dried.* Prepare a 1 mg/ml stock solution of the kanamycin sulfate working standard in solution 3 and store refrigerated for not longer than 1 month. Further dilute with solution 3.

Dose levels: 0.64, 0.80, 1.00, 1.25, and 1.56 μg/ml

Incubation: 36° – 37.5°C

9.61 Methicillin

Organism: *Staphylococcus aureus* ATCC 6538P
Maintain and prepare a suspension as described in Section 9.31. Dilute 1:20 and store refrigerated for not longer than 1 week.
Inoculum: 0.3 ml per 100 ml of medium 1
Medium: Base layer, 21 ml of medium 2. Seed layer, 4 ml of medium 1.
Standard: *Not dried.* Prepare a 1 mg/ml stock solution in solution 1 and store refrigerated for not longer than 4 days. Further dilute with solution 1.
Dose levels: 6.4, 8.00, 10.0, 12.5, and 15.6 μg/ml
Incubation: 32°–35°C

9.62 Mitomycin

Organism: *Bacillus subtilis* ATCC 6633
Prepare, standardize, and store the spore suspension as described in Section 9.55.
Inoculum: 0.5 ml per 100 ml of medium 8
Medium: Base layer, 10 ml of uninoculated medium 8. Seed layer, 4 ml of medium 8.
Standard: *Not dried.* Prepare a 1 mg/ml stock solution in solution 1 and store refrigerated for not longer than 14 days. Further dilute with solution 1.
Dose levels: 0.50, 0.71, 1.0, 1.41, and 2.0 μg/ml
Incubation: 36°–37.5°C

9.63 Nafcillin

Organism: *Staphylococcus aureus* ATCC 6538P
Maintain and prepare a suspension as described in Section 9.31. Dilute 1:20 and store refrigerated for not longer than 1 week.
Inoculum: 0.3 ml per 100 ml of medium 1
Medium: Base layer, 21 ml of medium 2. Seed layer, 4 ml of medium 1.
Standard: *Not dried.* Prepare a 1 mg/ml stock solution in solution 1 and store refrigerated for not longer than 2 days. Further dilute with solution 1.
Dose levels: 1.28, 1.60, 2.00, 2.50, and 3.12 μg/ml
Incubation: 32°–35°C

9.64 Natamycin

Organism: *Saccharomyces cerevisiae* ATCC 9763
Maintain and prepare the suspension as described in Section 9.33. Dilute 1:30 and store refrigerated for not longer than 4 weeks.
Inoculum: 0.8 ml per 100 ml of medium 19
Medium: *No base layer.* Seed layer, 8 ml of medium 19.
Standard: *Not dried.* Prepare a 1 mg/ml stock solution in dimethylformamide (DMF), further dilute with DMF to concentrations of 64, 80, 100, 125, and 156 μg/ml, and finally dilute each of these solutions 1:20 (e.g., 5 ml → 100 ml) with solution 10. Use the solutions the same day.
Dose levels: 3.20, 4.00, 5.00, 6.25, and 7.81 μg/ml

Incubation: 29° – 31°C

Remarks: Prepare the standard response line simultaneously with the sample solution using red low-actinic glassware. Use solutions within 2 hours of preparation.

9.65   Neomycin

9.65.1   Method 1

Organism: *Staphylococcus aureus* ATCC 6538P
Maintain and prepare a suspension as described in Section 9.31. Dilute 1 : 20 and store refrigerated for not longer than 1 week.
Inoculum: 0.4 ml per 100 ml of medium 11
Medium: Base layer, 21 ml of uninoculated medium 11. Seed layer, 4 ml of medium 11.
Standard: Proceed as described in Section 9.32. Prepare a 1 mg/ml stock solution in solution 3 and store refrigerated for not longer than 2 weeks. Further dilute with solution 3.
Dose levels: 6.4, 8.0, 10.0, 12.5, and 15.6 µg/ml
Incubation: 32° – 35°C

9.65.2   Method 2

Organism: *Staphylococcus epidermidis* ATCC 12228
Maintain and prepare a suspension as described in Section 9.31. Dilute 1 : 14 and store refrigerated for not longer than 1 week.
Inoculum: 1.0 ml per 100 ml of medium 11
Medium: As in method 1 (Section 9.65.1)
Standard: Proceed as in method 1 (Section 9.65.1).
Dose levels: 0.64, 0.80, 1.00, 1.25, and 1.56 µg/ml
Incubation: 36° – 37.5°C

9.66   Netilmicin

Organism: *Staphylococcus epidermidis* ATCC 12228
Maintain and prepare a suspension as described in Section 9.31. Dilute 1 : 14 and store refrigerated for not longer than 1 week.
Inoculum: 0.25 ml per 100 ml of medium 11
Medium: Base layer, 20 ml of uninoculated medium 11. Seed layer, 5 ml of medium 11.
Standard: *Not dried.* Prepare a 1 mg/ml stock solution in solution 3 and store refrigerated for not longer than 7 days. Further dilute with solution 3.
Dose levels: 0.064, 0.080, 0.100, 0.125, and 0.156 µg/ml
Incubation: 36° – 37.5°C
Remarks: Store the working standard below − 10°C under an atmosphere of nitrogen. Netilmicin sulfate is hygroscopic; therefore, exercise extra care during weighing.

9.67   Novobiocin

Organism: *Staphylococcus epidermidis* ATCC 12228
Maintain and prepare a suspension as described in Section 9.31. Dilute 1 : 14 and store refrigerated for not longer than 1 week.

Inoculum: 4.0 ml per 100 ml of medium 1
Medium: Base layer, 21 ml of medium 2. Seed layer, 4 ml of medium 1.
Standard: Proceed as described in Section 9.32, except dry the working standard at 100°C for 4 hours. Prepare a 10,000 μg/ml stock solution in absolute ethanol, and store refrigerated for not longer than 5 days. Dilute to 1 mg/ml with solution 3 and further dilute with solution 6.
Dose levels: 0.320, 0.400, 0.500, 0.625, and 0.781 μg/ml
Incubation: 34° – 36°C

9.68   Nystatin

Organism: Saccharomyces cerevisiae ATCC 2601
Maintain and prepare a suspension as described in Section 9.33. Dilute 1 : 30 and store refrigerated for not longer than 4 weeks.
Inoculum: 1.0 ml per 100 ml of medium 19
Medium: No base layer. Seed layer, 8 ml of medium 19.
Standard: Proceed as described in Section 9.32, except dry the working standard at 40°C for 2 hours. Prepare a 1000 U/ml stock solution in dimethylformamide (DMF), further dilute with DMF to concentrations of 256, 320, 400, 500, and 624 U/ml, and finally dilute each of these solutions 1 : 20 (e.g., 5 ml → 100 ml) with solution 6. *Use the same day.*
Dose levels: 12.8, 16.0, 20.0, 25.0, and 31.2 U/ml
Incubation: 29° – 31°C
Remarks: Prepare the standard response line simultaneously with the sample solution using red low-actinic glassware.

9.69   Oleandomycin

Organism: Staphylococcus epidermidis ATCC 12228
Maintain and prepare a suspension as described in Section 9.31. Dilute 1 : 14 and store refrigerated for not longer than 1 week.
Inoculum: 1.0 ml per 100 ml of medium 11
Medium: Base layer, 21 ml of uninoculated medium 11. Seed layer, 4 ml of medium 11.
Standard: Not dried. Prepare a 10,000 μg/ml stock solution in ethanol and store refrigerated for not longer than 30 days. Further dilute with solution 3.
Dose levels: 3.20, 4.00, 5.00, 6.25, and 7.81 μg/ml
Incubation: 36° – 37.5°C

9.70   Oxacillin

Organism: Staphylococcus aureus ATCC 6538P
Maintain and prepare a suspension as described in Section 9.31. Dilute 1 : 20 and store refrigerated for not longer than 1 week.
Inoculum: 0.3 ml per 100 ml of medium 1
Medium: Base layer, 21 ml of medium 2. Seed layer, 4 ml of medium 1.
Standard: Not dried. Prepare a 1 mg/ml stock solution in solution 1 and store refrigerated for not longer than 3 days. Further dilute with solution 1.

Dose levels: 3.20, 4.00, 5.00, 6.25, and 7.81 µg/ml
Incubation: 32° – 35°C

9.71 Paromomycin

Organism: Staphylococcus epidermidis ATCC 12228
Maintain and prepare a suspension as described in Section 9.31. Dilute 1:25 and store refrigerated for not longer than 1 week.
Inoculum: 2.0 ml per 100 ml of medium 11
Medium: Base layer, 21 ml of uninoculated medium 11. Seed layer, 4 ml of medium 11.
Standard: Proceed as described in Section 9.32. Prepare a 1 mg/ml stock solution in solution 3 and store refrigerated for not longer than 3 weeks. Further dilute with solution 3.
Dose levels: 0.64, 0.80, 1.00, 1.25, and 1.56 µg/ml
Incubation: 36° – 37.5°C

9.72 Penicillin G

Organism: Staphylococcus aureus ATCC 6538P
Maintain and prepare a suspension as described in Section 9.31. Dilute 1:20 and store refrigerated for not longer than 1 week.
Inoculum: 1.0 ml per 100 ml of medium 1
Medium: Base layer, 21 ml of medium 2. Seed layer, 4 ml of medium 1.
Standard: Not dried. Prepare a 1000 U/ml stock solution in solution 1 and store refrigerated for not longer than 4 days. Further dilute with solution 1.
Dose levels: 0.64, 0.80, 1.00, 1.25, and 1.56 U/ml
Incubation: 32° – 35°C

9.73 Penicillin V

Organism: Staphylococcus aureus ATCC 6538P
Maintain and prepare a suspension as described in Section 9.31. Dilute 1:20 and store refrigerated for not longer than 1 week.
Inoculum: 1.0 ml per 100 ml of medium 1
Medium: Base layer, 21 ml of medium 2. Seed layer, 4 ml of medium 1.
Standard: Not dried. Prepare a 100 IU/ml solution of penicillin V potassium working standard in solution 1 and store refrigerated for not longer than 4 days. Further dilute with solution 1.
Dose levels: 0.64, 0.80, 1.00, 1.25, and 1.56 U/ml
Incubation: 32° – 35°C

9.74 Plicamycin

Organism: Staphylococcus aureus ATCC 6538P
Maintain and prepare a suspension as described in Section 9.31. Dilute 1:20 and store refrigerated for not longer than 1 week.
Inoculum: 0.1 ml per 100 ml of medium 8
Medium: Base layer, 10 ml of uninoculated medium 8. Seed layer, 4 ml of medium 8.

Standard: Proceed as described in Section 9.32, except dry the working standard at 25°C for 4 hours. Prepare a 0.1 mg/ml solution in distilled water and store refrigerated for not longer than 1 day. Further dilute with solution 1.

Dose levels: 0.50, 0.70, 1.00, 1.41, and 2.00 μg/ml
Incubation: 32°–35°C

9.75 Polymyxin B

Organism: *Bordetella bronchiseptica* ATCC 4617
Maintain and prepare a suspension as described in Section 9.31. Dilute 1:20 and store refrigerated for not longer than 2 weeks.

Inoculum: 0.1 ml per 100 ml of medium 10
Medium: Base layer, 21 ml of medium 9. Seed layer, 4 ml of medium 10.
Standard: Proceed as described in Section 9.32. Dissolve by adding 2 ml of distilled water for each 5 mg of the working standard, then dilute with solution 6 to produce a stock solution of 10,000 U/ml. Store refrigerated for not longer than 2 weeks. Further dilute with solution 6.

Dose levels: 6.4, 8.00, 10.0, 12.5, and 15.6 U/ml
Incubation: 36°–37.5°C

9.76 Rifampin

Organism: *Bacillus subtilis* ATCC 6633
Maintain and prepare a suspension as described in Section 9.31. Store the standardized suspension for not longer than 6 months.

Inoculum: 0.1 ml per 100 ml of medium 2
Medium: Base layer, 21 ml of uninoculated medium 2. Seed layer, 4 ml of medium 2.
Standard: *Not dried.* Prepare a 1 mg/ml stock solution in methanol and store refrigerated for not longer than 1 day. Further dilute with solution 1.

Dose levels: 3.20, 4.00, 5.00, 6.25, and 7.81 μg/ml
Incubation: 29°–31°C

9.77 Sisomicin

Organism: *Staphylococcus epidermidis* ATCC 12228
Maintain and prepare a suspension as described in Section 9.31. Dilute 1:14 and store refrigerated for not longer than 1 week.

Inoculum: 0.03 ml per 100 ml of medium 11
Medium: Base layer, 21 ml of uninoculated medium 11. Seed layer, 4 ml of medium 11.
Standard: *Not dried.* Store the working standard below −20°C under an atmosphere of nitrogen. *Caution:* Sisomicin is hygroscopic. Add 2 ml of distilled water for each 5 mg of standard, prepare a 1 mg/ml stock solution in solution 3, and store refrigerated for not longer than 14 days. Further dilute with solution 3.

Dose levels: 0.064, 0.080, 0.100, 0.125, and 0.156 μg/ml
Incubation: 36°–37.5°C

9.78 Streptomycin

Organism: *Bacillus subtilis* ATCC 6633
Prepare, standardize, and store the spore suspension as described in Section 9.55.
Inoculum: Use test plates to determine the amount of inoculum needed for each 100 ml of medium 5.
Medium: Base layer, 21 ml of uninoculated medium 5. Seed layer, 4 ml of medium 5.
Standard: Proceed as described in Section 9.32. Prepare a 1 mg/ml stock solution in solution 3 and store refrigerated for not longer than 30 days. Further dilute with solution 3.
Dose levels: 0.64, 0.80, 1.00, 1.25, and 1.56 μg/ml
Incubation: 36° – 37.5°C

9.79 Ticarcillin

Organism: *Pseudomonas aeruginosa* ATCC 29336
Maintain by weekly subculturing on slants of medium 36 and incubate at 37°C for 24 hours. Using 3 ml of sterile USP saline TS, wash the growth from the slant onto a large agar surface, such as a Roux bottle, containing 250 ml of medium 36. Spread the organism over the entire surface, using sterile glass beads if necessary, and incubate the Roux bottle at 37°C for 24 hours. Wash the resulting growth from the agar surface with 50 ml of medium 37. Dilute 1 : 50 and store refrigerated for not longer than 1 week.
Inoculum: 1.5 ml per 100 ml of medium 38
Medium: Base layer, 21 ml of uninoculated medium 38. Seed layer, 4 ml of medium 38.
Standard: *Not dried.* Prepare a 1 mg/ml stock solution in solution 1 and store refrigerated for not longer than 1 day. Further dilute with solution 1.
Dose levels: 3.20, 4.00, 5.00, 6.25, and 7.81 μg/ml
Incubation: 36° – 37.5°C

9.80 Vancomycin

Organism: *Bacillus subtilis* ATCC 6633
Prepare, standardize, and store the spore suspension as described in Section 9.55.
Inoculum: Use test plates to determine the amount of inoculum required for each 100 ml of medium 8.
Medium: Base layer, 10 ml of uninoculated medium 8. Seed layer, 4 ml of medium 8.
Standard: Proceed as described in Section 9.32. Prepare a 1 mg/ml stock solution in distilled water and store refrigerated for not longer than 1 week. Further dilute with solution 4.
Dose levels: 6.4, 8.0, 10.0, 12.5, and 15.6 μg/ml
Incubation: 36° – 37.5°C

References

Andrew, M. L., and Weiss, P. J. (1959). *Antibiot. Chemother.* **9**, 277.
Arret, B., Woodard, M. R., Wintermere, D. M., and Kirshbaum, A. (1957). *Antibiot. Chemother.* **7**, 545.

Kavanagh, F. W. (1963). "Analytical Microbiology," Vol. I. Academic Press, New York and London.

Shaw, W. H. C., and Vincent, S. (1972). *In* "Analytical Microbiology" (F. W. Kavanagh, ed.), Vol. II. Academic Press, New York and London.

United States Code of Federal Regulations (1985). 21 CFR 436.100. The Office of the Federal Register, National Archives and Records Administration, Washington, D.C.

Vincent, S., and Shaw, W. H. C. (1972). *In* "Analytical Microbiology" (F. W. Kavanagh, ed.), Vol. II. Academic Press, New York and London.

Weiss, P. G., Andrew, M. L., and Wright, W. W. (1957). *Antibiot. Chemother.* 7, 374.

CHAPTER 10

DIFFUSION ASSAY METHODS FOR VITAMINS

10.1 Introduction

Some very good reference books which contain assay methods for water-soluble vitamins include those by György (1950), Barton-Wright (1952), and Kavanagh (1963, 1972).

The diffusion assay methods for individual vitamins presented in this chapter have been much simplified to eliminate the time-consuming media preparation and tedious lengthy centrifugation of cultures. Complete-formula, commercially available, dehydrated vitamin assay media replace the difficult preparation of complex media by a simple weighing and solubilization process. Simply follow the instructions on the label of the bottle which normally produces double-strength medium. The authors have found the dehydrated media supplied by Difco Laboratories satisfactory.

The preparation of inoculum has also been simplified by adopting the vitamin depletion method described by Gare (1968). Most of the vitamins are available as pure chemicals, which makes the selection and establishment of standards a simple operation. Samples containing mixtures of vitamins such as syrups and other aqueous vitamin solutions require only dilutions in distilled water, making sample preparation quite simple. Certain samples may need extraction by hydrolysis or enzymatic digestion, especially samples of natural plant or animal origin, in which case the above-mentioned reference books should be consulted.

The same basic techniques are used for the vitamin assays as for the assay of antibiotics. Therefore, once a diffusion assay laboratory has been established, the same equipment can be used for assaying vitamins as well. The guidelines recommended for the individual vitamin assays have to be followed together with the fuller details given in earlier chapters. Furthermore, the use of large plates is assumed for the vitamin assays because of their obvious advantages, although the methods recommended here can be easily adapted to petri dish–cylinder plate assays if desired. Two-dose level assays are usually quite satisfactory; however, these can be easily increased to three- or four-dose levels, provided the same dose ratios are maintained.

10.2 Thiamine

Other names: Aneurin hydrochloride, vitamin B_1

Test organism: *Lactobacillus fermentum* NCIB 6991, ATCC 9338

Maintain by weekly transfer alternately on microassay culture agar stabs and microinoculum broth (Difco), as described in Chapter 2. Incubate at 37°C for 48 hours.

Inoculum medium: Thiamine assay medium, Difco

Prepare the medium single strength. (*Note:* Instructions on the label refer to double strength.) Fill 100-ml amounts into 500-ml medical flat or Roux bottles, and sterilize at 121°C for 15 minutes.

Preparation of inoculum: Suspend the growth of a 24-hour stab culture in about 5 ml of inoculum medium taken from one of the bottles, and return the suspension back into the same bottle. Incubate the inoculated inoculum medium at 37°C for 18–24 hours with the bottle lying on its side to allow maximum surface area and with the cap loosened. If not used immediately after incubation, store in the refrigerator for not longer than 24 hours.

Assay media: Prepare separately.

(1) Double-strength assay medium

Use thiamine assay medium, Difco. For each large plate prepare 115 ml of double-strength medium according to the instructions on the label.

(2) Double-strength agar

By means of a small measure, transfer about 3.2 g of powdered agar to each 115 ml of distilled water in a suitable large bottle, such as a Roux bottle or a 300-ml medical flat bottle. (*Note:* Ensure that the agar is free from trace vitamins.) Allow the agar to soak in the water for not less than 15 minutes, and sterilize at 121°C for 20 minutes. Store at room temperature until required.

Preparation of plates: Melt the double-strength agar for about 45–60 minutes. Allow to cool to below 80°C, then transfer it to a water bath set at 50°C. Prepare the double-strength assay medium, pour it into the bottle containing the double-strength agar, and allow the temperature of the combined medium to drop just below 50°C. Shake the bottle of inoculum to produce a uniform suspension and mix 10 ml with the combined agar medium. Pour the seeded medium *immediately* into a previously leveled large plate, allow to solidify, and then proceed as described in the general chapters.

Preparation of standard solution: Accurately weigh about 100 mg of the working standard previously dried at 100°C for 3 hours and cooled in a desiccator over phosphorus pentoxide. Quantitatively transfer to a 1000-ml volumetric flask, dissolve, and dilute to volume with 2% (v/v) hydrochloric acid in glass-distilled water. Store this stock solution in the refrigerator for not longer than 14 days. Prepare daily working standard solutions of 2.0 and 1.0 μg/ml by diluting the primary stock solution with glass-distilled water. Divide the actual standard concentration obtained by the theoretical concentration to calculate the factor of the standard. For example,

$$\frac{2.03}{2.00} = 1.015$$

Preparation of sample solutions: Dilute solutions to plating-out levels of 2.0 and 1.0 μg/ml using distilled water. Prepare solids in the same way as the standard.

Incubation: After plating out the solutions using statistical designs as described in previous chapters, incubate plates at 37°C for 16–18 hours.

Note: The original assay was developed by Bacharach and Cuthbertson (1948).

10.3 Riboflavin

Other names: Riboflavine, vitamin B_2
Test organism: *Lactobacillus casei* subsp. *rhamnosus* NCIB 8010, ATCC 7469
Maintain by weekly transfer alternately on microassay culture agar stabs and microinoculum broth (Difco), as described in Chapter 2. Incubate at 37°C for 18 hours, and store in the refrigerator.
Inoculum media: Medium A, single strength
Prepare riboflavin assay medium (Difco), single strength. (*Note:* Instructions on the bottle label are for double strength.) Add sufficient riboflavin solution to the medium to contain 1.0 μg riboflavin per milliliter. Fill in 100-ml amounts into 500-ml medical flat or Roux bottles, and sterilize by autoclaving at 121°C for 15 minutes.

Medium B, single strength
Prepare riboflavin assay medium (Difco) single strength. (*Note:* Instructions on the bottle label refer to double strength.) This medium *must not contain riboflavin.* Fill in 100-ml amounts, and sterilize as inoculum medium A.
Preparation of inoculum: Add 2–3 ml of inoculum medium A to an overnight stab culture of *L. casei,* and emulsify the growth by using a sterile bacteriological straight wire. Return the suspension back into the same bottle of inoculum medium A and incubate for 16–18 hours at 37°C, placing the bottle on its side and with the cap loosened. The next day, tighten the cap and shake the bottle to produce a uniform suspension; then aseptically transfer about 10 ml of this growth from medium A to 100 ml of inoculum medium B. Again incubate as above at 37°C for a further 16–18 hours. Use 10 ml of this seeding inoculum for each bottle of the combined assay medium as described below.
Assay media: Prepare separately.

(1) Double-strength assay medium
For each large plate prepare 115 ml of riboflavin assay medium (Difco) according to the instructions on the label.
(2) Double-strength agar
Prepare as described in Section 10.2.
Preparation of plates: For each large plate, melt one bottle of the double-strength agar by steaming for 45–60 minutes. Place this, after cooling to below 80°C, together with one bottle of the double-strength assay medium in a water bath set at 50°C and allow the temperature of each to come to 50°C. Mix the two bottles of media, add 10 ml of the seeding inoculum (see under "Preparation of Inoculum"), and mix again. Pour the seeded medium *immediately* into a previously leveled large plate, and proceed as described in the general chapters.
Preparation of standard solution:

Note: Protect from light throughout all operations.

Accurately weigh about 100 mg of the working standard, dry at 100°C for 3 hours, and allow to cool in a desiccator over phosphorus pentoxide. Reweigh and quantitatively transfer to a beaker containing about 30 ml of distilled water. Bring to the boil, then add 5 ml of glacial acetic acid. (*Safety note:* Use safety pipette and eye protection.) Stir with a glass rod over a gentle flame until dissolved. Allow to cool to room temperature, quantitatively transfer to a 1000-ml volumetric flask, and dilute to volume with distilled water. Store this stock solution in the refrigerator for not longer than 14 days.
Prepare daily working standard solutions of approximately 1.0 and 0.5 μg/ml by diluting the

primary stock solution with distilled water. Calculate the factor by which these solutions differ from the theoretical dose levels by dividing the actual plating-out concentration obtained on the day of the assay by the theoretical level, for example:

$$\frac{1.096}{1.000} = 1.096$$

Preparation of sample solutions: Dilute sample solutions to plating-out levels of 1.0 and 0.5 μg/ml using distilled water. (*Note:* Protect from light.)

Incubation: 37°C for 16–18 hours

10.4 Pyridoxine

Other names: Vitamin B_6, pyridoxal, pyridoxamine

Test organism: Saccharomyces uvarum (carlsbergensis) NCTC 4228, ATCC 9080

Maintain by monthly transfer on Sabouraud agar slants and incubating at 30°C for 48 hours. Store in the refrigerator after incubation.

Preparation of inoculum: For each large plate, emulsify the growth of one slant culture of *S. uvarum* in 10 ml of sterile normal saline. Add this inoculum to each bottle of the combined assay medium at 50°C. (See below.)

Assay media: Prepare separately.

(1) Double-strength assay medium

Prepare pyridoxine-Y medium (Difco), according to the instructions on the label.

(2) Double-strength agar

Prepare as described in Section 10.2.

Preparation of plates: Proceed as described in Section 10.3.

Preparation of standard solutions: Dry about 100 mg of the working standard at 100°C for 3 hours, allow to cool to room temperature in a desiccator over phosphorus pentoxide, and reweigh. Quantitatively transfer to a 1000-ml volumetric flask, dissolve, and make up to volume with distilled water. Store in the refrigerator for not longer than 14 days. Prepare daily working solutions by diluting the primary stock solution of the working standard with distilled water to dose levels of 0.2 and 0.1 μg/ml. Calculate the factor by which these solutions differ from theory; for example:

$$\frac{0.198 \ \mu g/ml}{0.200 \ \mu g/ml} = 0.99$$

Preparation of sample solutions: Prepare dilutions in distilled water to expected dose levels of 0.2 and 0.1 μg/ml.

Incubation: 30°C for 16–18 hours

Note: The original assay was developed by Jones and Morris (1950).

10.5 Cyanocobalamin

Other name: Vitamin B_{12}

Test organism: Escherichia coli M200 NCIB 9270, ATCC 14169

Maintain by weekly transfers on the culture maintenance medium given below. Incubate at 27°C for 18–24 hours and store in the refrigerator until required.

Inoculum media:
Medium A, culture maintenance medium

Acid-hydrolyzed casein	6.0 g
Dipotassium hydrogen phosphate (K_2HPO_4)	0.2 g
Ferrous sulfate ($FeSO_4 \cdot 7H_2O$)	0.005 g
Magnesium sulfate ($MgSO_4 \cdot 7H_2O$)	0.2 g
L-asparagine	0.15 g
Glass-distilled water (after further ingredients) to	1000.0 ml

Dissolve the constituents one by one with gentle heating in the above order, adding a few drops of concentrated hydrochloric acid to dissolve the asparagine, and adjust the pH of the solution to 7.2 ± 0.1, boil, and filter. After cooling, add 2.0 g of glycerol and 15 g of bacteriological agar. Make up the volume to 1 liter with glass-distilled water, and autoclave at 121°C for 20 minutes to dissolve the agar. Readjust the pH to 7.2 ± 0.1 and add the equivalent of 400 μg of cyanocobalamin. After thorough mixing, dispense the medium in 10-ml amounts into suitable containers such as 25-ml universal screw-capped bottles. Sterilize at 121°C for 15 minutes, and after sterilization allow the medium to set in an inclined position to present maximum surface area. Store in a refrigerator for not longer than 6 months.

Medium B, inoculum medium
Dissolve 20 g of bacteriological peptone and 5 g of sodium chloride in glass-distilled water, and make up the volume to 2 liters. Adjust the pH to 7.2 ± 0.1, and dispense in flasks or bottles so that 100 ml in each occupies a depth of between 3 and 5 mm. Autoclave the medium at 121°C for 15 minutes, and store in the refrigerator until required after testing for sterility. (This is done by incubating the autoclaved bottles at 30°–35°C for 5–7 days.)

Preparation of inoculum: Using a bacteriological loop, inoculate a bottle of inoculum medium from a freshly grown slant culture and incubate at 27°C for 18–24 hours with the bottle lying on its side to allow a maximum surface area and with the cap loosened. Using a sterile graduated pipette, transfer 15 ml of the incubated culture to about 100 ml of sterile physiological salt solution and mix. Use 1 ml of this dilute inoculum for each bottle of the combined assay medium.

Assay media:
(1) Double-strength assay medium

Dipotassium hydrogen phosphate (K_2HPO_4)	140 g
Potassium dihydrogen phosphate (KH_2PO_4)	60 g
Sodium citrate ($Na_3C_6H_5O_7 \cdot 2H_2O$)	10 g
Magnesium sulfate ($MgSO_4 \cdot 7H_2O$)	2 g
Ammonium sulfate [$(NH_4)_2SO_4$]	20 g
Sodium chloride (NaCl)	1 g
Distilled water to	10 liters

Dissolve the ingredients separately to prevent precipitation. Mix the solutions and adjust the pH to 7.0. Dilute to volume and filter if necessary. Dispense in 125-ml amounts into 500-ml medical flats or Roux bottles, preferably fitted with screw caps. Sterilize at 121°C for 20 minutes, and after cooling, store in the refrigerator.

(2) Double-strength agar
Prepare as described in Section 10.2.

(3) Dextrose solution
Prepare a 40% (w/v) solution of dextrose AR in distilled water. Sterilize at 121°C for 15 minutes, and after cooling, store in the refrigerator.

Preparation of plates: Melt the double-strength agar by steaming for 45–60 minutes.

When the temperature has dropped below 80°C, transfer into a water bath set at 50°C. Place the double-strength assay medium also in the water bath, and when its temperature has reached 50°C, mix with the agar. To each bottle of combined assay medium, add 1.5 ml of the sterile 40% dextrose solution. Cool the medium to 47°–48°C, add 1.0 ml of the dilute inoculum to each bottle of medium, mix and pour *immediately* into previously leveled large plates, then proceed as described in the general chapters.

Preparation of standard solutions:

(1) Standardization

Standardize a solution of cyanocobalamin (e.g., Cytamen 250, Glaxo) spectrophotometrically and also by microbiological assay against a previously verified standard. Use the average potency obtained from determinations carried out by a minimum of two different operators on each of 3 days.

(2) Primary stock solution

Dilute 1.0 ml of standardized solution to about 70 ml with distilled water, add one drop of 0.1% (w/v) aqueous potassium cyanide. (*WARNING:* Potassium cyanide is poisonous. Use safety pipette.) Make up volume to 100 ml using distilled water. Allow to stand at room temperature for at least 10 minutes before further dilution. Protect from light and store in the refrigerator for not longer than 7 days.

(3) Dose levels

Prepare fresh working standards daily as required by diluting aliquots of the primary stock solution to plating-out levels of 0.01 μg/ml and 0.0025 μg/ml using distilled water. Calculate the factor of the standard by dividing the actual plating-out concentration by the theoretical level, for example:

$$\frac{0.0086 \ \mu\text{g/ml}}{0.0100 \ \mu\text{g/ml}} = 0.860$$

Preparation of sample solutions: Dilute the samples according to their estimated potency to concentrations of 0.01 and 0.0025 μg/ml using distilled water. Add one drop of 0.1% (w/v) aqueous potassium cyanide to the primary flask. (*WARNING:* Potassium cyanide is poisonous. Use safety pipette.) Allow to stand at room temperature for 10 minutes before further dilution.

Incubation: 27°C for 16–18 hours

Remarks: Other cobalamins such as hydroxocobalamin and methylcobalamin can also be assayed by the same method, but the relevant standard must be used because the diffusion properties are different, and the potassium cyanide treatment of both the standard and sample solutions must be omitted.

10.6   Biotin

Other names: Vitamin H, coenzyme R

Test organism: *Lactobacillus plantarum* NCIB 8030, ATCC 8014

Maintain by weekly transfer alternately on microassay culture agar stabs and microinoculum broth (both Difco), as described in Chapter 2. After incubation at 30°C for 18 hours, store all cultures in the refrigerator.

Inoculum media:

Medium A, single strength

Prepare medium A as described in Section 10.3.

Medium B, single strength
Prepare single-strength biotin assay medium, Difco. (*Note:* Instructions on the label are for double-strength medium.) Fill in 100-ml amounts into a Roux or a 500-ml medical flat bottle with screw caps. Sterilize at 121°C for 15 minutes, and store in the refrigerator.

Preparation of inoculum: Add 2–3 ml sterile normal saline to a freshly grown stab culture, and emulsify the growth with the aid of a sterile bacteriological straight wire. Transfer the suspension into inoculum medium A, and incubate for 16–18 hours at 30°C with the bottle lying flat and the cap loosened. The next day, transfer about 20 ml of this growth from medium A to 100 ml of medium B, and again incubate at 30°C for 16–18 hours. Use about 10 ml of this inoculum for each bottle of the combined assay medium.

Assay media: Prepare separately.
(1) Double-strength assay medium
For each large plate, prepare 115 ml of double-strength biotin assay medium, Difco.
(2) Double-stength agar
Prepare as described in Section 10.2.

Preparation of plates: For each large plate, melt one bottle of double-strength agar by steaming for 45–60 minutes. After cooling to below 80°C, place the agar together with one bottle of the double-strength assay medium in a water bath at 50°C and allow the temperature of each to reach 50°C. Mix the two bottles of media, add 10 ml of the inoculum, and mix again. Pour the combined inoculated medium *immediately* into the previously leveled plate, and proceed as described in the general chapters.

Preparation of standard solution: Accurately weigh about 50 mg of the working standard and dissolve in sufficient distilled water to produce a solution containing about 1.0 μg/ml. Store the primary stock solution in the refrigerator for not longer than 1 month. Prepare daily plating-out solutions in distilled water containing approximately 0.05 and 0.025 μg/ml. Calculate the factor of the standard by dividing the actual plating-out concentration by the theoretical dose level, for example:

$$\text{Factor of standard} = \frac{0.0532 \ \mu\text{g/ml}}{0.0500 \ \mu\text{g/ml}} = 1.064$$

Preparation of sample solutions: Dilute samples according to their estimated potency to concentrations of 0.05 and 0.025 μg/ml using distilled water.
Incubation: 30°C for 16–18 hours

10.7 Folic Acid

Other names: Folacin, pteroylglutamic acid
Test organism: *Streptococcus faecalis* NCIB 6459, ATCC 8043
Maintain by weekly transfer alternately on microassay culture agar stabs and microinoculum broth (Difco), as described in Chapter 2. Incubate at 37°C and store in the refrigerator.
Inoculum media:
Medium A, liver tryptone broth

Glucose	0.2 g
Tryptone, Difco	1.0 g
Potassium phosphate (K_2HPO_4)	0.5 g
Yeast extract, Difco	0.2 g
Liver extract	5.0 ml
Distilled water to	200.0 ml

Note: Add liver extract *after* dissolving other ingredients.

Fill in 10-ml amounts into 25-ml universal bottles with screw caps, and sterilize at 121 °C for 10 minutes.

Medium B, single-strength inoculum medium
Prepare single-strength folic acid AOAC medium, Difco. (*Note:* Instructions on the label are for double-strength medium.) Fill in 80-ml amounts into Roux bottles or 500-ml medical flats with screw caps, and sterilize at 121 °C for 15 minutes.

Preparation of inoculum: Wash off the growth from two recent stab cultures with the aid of a bacteriological straight wire into 10 ml of inoculum medium A, and incubate at 37 °C overnight. Mix to obtain a uniform suspension, and transfer 0.25 ml of the overnight culture into each of three further 10 ml inoculum medium A, and again incubate at 37 °C overnight. The next morning, transfer the contents of all three bottles into 80 ml of inoculum medium B and incubate at 37 °C for about 24 hours. Use about 20 ml of this culture to inoculate each bottle of the combined assay medium as described below.

Assay media:
(1) Double-strength assay medium
For each large plate, prepare about 120 ml of folic acid AOAC assay medium as instructed on the label. Do not store the medium but prepare fresh.
(2) Double-strength agar
Prepare as described in Section 10.2.

Preparation of plates: As soon as the folic acid double-strength medium begins to boil, allow to cool to below 80 °C and add it to a bottle of double-strength agar which was previously melted and cooled to about 50 °C. Mix and allow to cool to 50 °C, then add 20 ml of inoculum and pour *immediately* into previously leveled plates. Proceed further as described in the general chapters.

Preparation of standard solutions: Suspend about 60 mg of pure folic acid working standard in about 30 ml of distilled water in a 1000-ml volumetric flask. Add five drops of concentrated ammonia solution to dissolve the folic acid and make up to volume with glass-distilled water. Store this stock solution in the refrigerator for not longer than 14 days.

Prepare daily working standards of 0.20 and 0.05 μg/ml by diluting the primary stock solution with glass-distilled water. Calculate the factor by which the actual concentrations differ from the theoretical dose levels, for example:

$$\text{Factor} = \frac{0.2026 \ \mu\text{g/ml}}{0.2000 \ \mu\text{g/ml}} = 1.013$$

Preparation of sample solutions: Dilute samples according to their estimated potency to plating-out levels of 0.20 and 0.05 μg/ml using distilled water.
Incubation: 37 °C for 16–18 hours

Note: For further details on the assay of folates refer to Kavanagh (1972), Chapter 5.

10.8   Nicotinic Acid

Other names: Niacin, vitamin PP
Test organism: *Lactobacillus plantarum* NCIB 8030, ATCC 8014

Maintain by weekly transfer alternately on microassay culture agar stabs and microinoculum broth (Difco), as described in Chapter 2. After incubation at 30°C for about 18 hours, store all cultures in the refrigerator.

Inoculum media:
Medium A, single strength
Prepare as described in Section 10.3.

Medium B, single strength
Prepare single-strength niacin assay medium (Difco). (*Note:* Label instructions are for double strength.) Fill 100-ml amounts into Roux bottles or 500-ml medical flats. Sterilize by autoclaving at 121°C for 15 minutes.

Preparation of inoculum: Emulsify growth of an 18-hour stab culture in about 5 ml of inoculum medium A, and transfer the suspension back into the same bottle of medium. Incubate at 37°C for 18–24 hours with the bottle lying on its side to allow maximum surface area, and with the cap loosened. The next day, transfer about 20 ml of this growth from medium A to 100 ml of medium B, and again incubate at 37°C for a further 16–18 hours.

Assay media:
(1) Double-strength assay medium
For each large plate, prepare 115 ml of double-strength assay medium according to the instructions on the label of niacin assay medium, Difco.

(2) Double-strength agar
Prepare as described in Section 10.2.

Preparation of plates: Melt the double-strength agar by steaming for 45–60 minutes and, after cooling to below 80°C, place in a water bath set at 50°C. In the meantime prepare the double-strength assay medium and mix it with the double-strength agar. Allow the temperature of the combined medium to reach 50°C, and mix with 10 ml of inoculum taken from inoculum medium B. *Immediately* pour the seeded medium into previously leveled plates, and proceed as described under the general chapters.

Preparation of standard solutions: Dry about 50 mg pure nicotinic acid working standard at 100°C for 3 hours, allow to cool in a desiccator over phosphorus pentoxide, and weigh again. Transfer quantitatively to a 1000-ml volumetric flask and dilute to volume with distilled water. Store the stock solution in the refrigerator for not longer than 14 days.

Prepare daily working solutions of approximately 5.0 and 2.5 µg/ml by diluting the primary stock solution with distilled water. Calculate the factor by which these solutions differ from their nominal strengths by dividing the actual plating-out concentration obtained by the theoretical level, for example:

$$\text{Factor} = \frac{5.329 \ \mu\text{g/ml}}{5.000 \ \mu\text{g/ml}} = 1.0658$$

Preparation of sample solutions: Prepare dilutions according to the assumed potency of the sample to plating-out levels of 5.0 and 2.5 µg/ml using distilled water.

Incubation: Incubate the plates at 30°C for 16–18 hours.

10.9 Pantothenate

Test organism: *Lactobacillus plantarum* NCIB 8030, ATCC 8014
Maintain by weekly transfer alternately on microassay culture agar stabs and microinocu-

lum broth (Difco), as described in Chapter 2. After incubation at 30°C for 18 hours, store all cultures in the refrigerator.

Inoculum media:

Medium A, single strength

Prepare microinoculum broth (Difco), according to the instructions on the label, and fill in 10-ml amounts into 25-ml universal bottles with screw caps. Sterilize at 121°C for 15 minutes.

Medium B, single strength

Prepare single-strength pantothenate assay medium, Difco. (*Note:* Label instructions are for double-strength medium.) Fill in 100-ml amounts into Roux bottles or 500-ml medical flats. Sterilize at 121°C for 10 minutes.

Preparation of inoculum: Three days before the intended assay, transfer about 0.25 ml of a recent broth culture into 10 ml of inoculum medium A and incubate at 37°C for 16–18 hours.

The following day, transfer the whole 10 ml of the above culture into 100 ml of inoculum medium B, and incubate at 37°C for 16–18 hours, placing the bottle on its side with the cap loosened.

On the third day (i.e., the day before the intended assay), transfer about 20 ml of the above culture into another 100 ml of inoculum medium B and again incubate as above at 37°C overnight. Use about 15 ml of this culture to inoculate one bottle of the combined assay medium.

Assay media:

(1) Double-strength assay medium

For each large plate, prepare about 100 ml of pantothenate assay medium (Difco), according to the instructions on the label.

(2) Double-strength agar

Prepare as described in Section 10.2.

Preparation of plates: Melt one bottle of the double-strength agar for each large plate by steaming for 45–60 minutes. When the temperature has dropped to below 80°C, place this together with one bottle of the double-strength assay medium in a water bath set at 48°C and allow the temperature of each to come to 48°C. Mix the two bottles of media, add 15 ml of the seeding inoculum, and mix again. *Immediately* pour the seeded medium into a previously leveled large plate, and proceed as described in the general chapters.

Preparation of standard solutions: Weigh about 100 mg of pure calcium pantothenate working standard and dry for 3 hours at 100°C. Allow to cool over phosphorus pentoxide in a desiccator, and reweigh. Quantitatively transfer into a 1000-ml volumetric flask, dissolve in glass-distilled water, and make up to volume with glass-distilled water. Store this stock solution in the refrigerator for not longer than 14 days.

Prepare daily working standard solutions of 2.0 and 1.0 μg/ml by diluting the stock solution with glass-distilled water. Calculate the factor by which these solutions differ from the theoretical levels by dividing the actual plating-out concentration obtained by the theoretical level, for example:

$$\text{Factor} = \frac{2.192 \ \mu g/ml}{2.000 \ \mu g/ml} = 1.096$$

Preparation of sample solutions: Dilute samples according to their assumed potency to plating-out levels of 2.0 and 1.0 μg/ml using glass-distilled water.

Incubation: 30°C for 16–18 hours

Note: Diffusion assay methods for biotin, nicotinic acid, and pantothenic acid were first described by Morris and Jones (1953).

References

Bacharach, A. L., and Cuthbertson, W. F. J. (1948). *Analyst* 73, 334.
Barton-Wright, E. C. (1952). "The Microbiological Assay of the Vitamin B Complex and Amino Acids." Pitman, London.
Gare, L. (1968). *Analyst,* 93, 456.
György, P., ed. (1950). "Vitamin Methods," Vol. 1. Academic Press, New York.
Jones, A., and Morris, S. (1950). *Analyst* 75, 608.
Kavanagh, F., ed. (1963). "Analytical Microbiology," Vol. I. Academic Press, New York.
Kavanagh, F., ed. (1972). "Analytical Microbiology," Vol. II. Academic Press, New York and London.
Morris, S., and Jones, A. (1953). *Analyst* 78, 15.

CHAPTER 11

PRACTICAL TUBE ASSAYS

The contents of this chapter are divided into two parts: Section 11A deals with the tube assay of individual antibiotics and Section 11B describes tube assays of vitamins.

11A ANTIBIOTICS

11.1 Introduction

The relatively simple principles of turbidimetric determination of growth-inhibiting substances were outlined in Chapter 5, together with the many pitfalls and problems associated with these assays in laboratory practice.

It is possible to carry out a limited number of antibiotic tube assays on a small scale by less sophisticated manual methods, but to control the many variables adequately requires an efficient laboratory organization, a lot of expertise, and careful attention to detail.

Compared to diffusion assays, even the simplest manual approach will require the purchase of some expensive equipment, such as a suitable spectrophotometer and a water bath which is capable of controlling the temperature to within $\pm 0.1\,°C$. From culture preparation to accurate sample dilution, from determining the critical narrow dose range to accurate timing of incubation, the whole process is full of pitfalls for the unwary beginner.

Once assay conditions have been standardized, a method has been shown to be reproducible, and sufficient experience has been gained, it is possible to obtain reliable results within a working day — which is one of the main advantages of this method compared to the diffusion assays. But to make the method work on a large scale it is necessary to introduce a fair degree of automation, which makes this method more suitable for a busy pharmaceutical assay laboratory because of the capital outlay. Only after a time will the automated system show a savings in the long run because of the high throughput of samples and a possible savings in staff. The most reliable and most highly developed automated system is the Autoturb, which was developed and described by Kavanagh and Kuzel (Kuzel and Kavanagh, 1971) (see also Section 5.6).

The tube assay of antibiotics has never gained popularity in the United Kingdom and consequently the authors have not gained as much experience in their use as in diffusion assays. The methods described in the following pages are based on the currently available official methods, which are summarized in Table 11.1. The methods that follow give only the bare essentials for these assays and must be supplemented by the fuller details in Chapter 5.

In all the assays that are outlined here, the dose levels approximate to a geometric progression. This may be illustrated by reference to the assay for amikacin, which has dose levels of 8.0, 8.9, 10.0, 11.2, and 12.5 μg/ml. These are approximations for 8.000, 8.944, 10.000, 11.180, and 12.500 μg/ml, which form a geometrical progression in which the ratio between adjacent dose levels is $(1.25)^{0.5}:1$ or $1.118034:1$. This same ratio is used in all the assays outlined here except that for candicidin, which employs a dose ratio of $(2)^{0.5}:1$ or $1.4142136:1$.

Formulas for media and diluents are listed in Appendix 4.

Table 11.1

Official Tube Assays of Antibiotics[a]

Assay	21 CFR	USPXXI	BP	EP
Amikacin	*	*	—	—
Candicidin	*	*	—	—
Capreomycin	*	*	—	—
Chloramphenicol	*	*	—	—
Chlortetracycline	*	*	*	*
Cycloserine	*	*	—	—
Demeclocycline	*	*	*	*
Dihydrostreptomycin	*	*	—	—
Doxycycline	*	*	—	—
Erythromycin	—	—	*	*
Gramicidin	*	*	—	—
Kanamycin	*	*	—	—
Lincomycin	*	—	—	—
Meclocycline	*	—	—	—
Methacycline	*	*	—	—
Minocycline	*	*	—	—
Neomycin	—	—	*	*
Oxytetracycline	*	*	*	*
Rifamycin sodium	—	—	*	*
Rolitetracycline	*	*	—	—
Spectinomycin	*	*	—	—
Streptomycin	*	*	*	*
Tetracycline	*	*	*	*
Tobramycin	*	*	*	—
Troleandomycin	*	*	—	—
Tyrothricin	*	—	—	—
Viomycin	—	*	—	—

[a] (*) Described; (−) not described.

11.2   Amikacin

Test organism: Staphylococcus aureus ATCC 6538P
Maintain the culture, prepare the inoculum, and store as described in Section 9.31.
Inoculum: Add 0.1 ml of the standardized inoculum to each 100 ml of medium 3.
Standard: Not dried. Prepare a 1 mg/ml stock solution in distilled water and store in a refrigerator for not longer than 2 weeks. Further dilute with distilled water.
Dose levels: 8.0, 8.9, 10.0, 11.2, and 12.5 μg/ml
Incubation: 36° – 37.5°C

11.3   Candicidin

Test organism: Saccharomyces cerevisiae ATCC 9763
Maintain the culture, prepare the inoculum, and store as described in Section 9.33.
Inoculum: Add 0.2 ml of the standardized suspension to each 100 ml of medium 13.
Standard: Prepare a 1 mg/ml stock solution in dimethyl sulfoxide after drying the working standard at 40°C and a pressure of 5 mm Hg or less for 3 hours. Use the solution only on the day of preparation, and further dilute with distilled water.
Dose levels: 0.030, 0.043, 0.060, 0.085, and 0.120 μg/ml
Incubation: 27° – 29°C

Note: Sterile equipment must be used throughout the assay.

11.4   Capreomycin

Test organism: Klebsiella pneumoniae ATCC 10031
Maintain the culture, prepare the suspension, and store as described in Section 9.31.
Inoculum: Add 0.05 ml of the standardized inoculum to each 100 ml of medium 3.
Standard: Proceed as described in Section 9.32, except dry the working standard at 100°C for 4 hours. Prepare a 1 mg/ml stock solution in distilled water, and store in the refrigerator for not longer than 7 days. Further dilute with distilled water.
Dose levels: 80.0, 89.0, 100.0, 112.0, and 125 μg/ml
Incubation: 36° – 37.5°C

11.5   Chloramphenicol

Test organism: Escherichia coli ATCC 10536
Maintain the culture, prepare the suspension, and store as described in Section 9.31.
Inoculum: Add 0.7 ml of the standardized suspension to each 100 ml of medium 3.
Standard: Not Dried. Prepare a 10 mg/ml stock solution in ethanol, dilute to 1 mg/ml with distilled water, and store in a refrigerator for not longer than 1 month. Further dilute with distilled water.
Dose levels: 2.00, 2.24, 2.50, 2.80, and 3.12 μg/ml
Incubation: 36° – 37.5°C

11.6   Chlortetracycline

Test organism: *Escherichia coli* ATCC 10536
Maintain the culture, prepare the suspension, and store as described in Section 9.31.
Inoculum: Add 0.1 ml of the standardized suspension to each 100 ml of medium 3.
Standard: Not dried. Prepare a 1 mg/ml stock solution in 0.01 *N* hydrochloric acid, and store in a refrigerator for not longer than 4 days. Further dilute with distilled water.
Dose levels: 0.048, 0.054, 0.060, 0.067, and 0.075 µg/ml
Incubation: 36° – 37.5°C

11.7   Cycloserine

Test organism: *Staphylococcus aureus* ATCC 6538P
Maintain the culture, prepare the suspension, and store as described in Section 9.31.
Inoculum: Add 0.4 ml of the standardized suspension to each 100 ml of medium 3.
Standard: Proceed as directed in Section 9.32. Prepare a 1 mg/ml stock solution in distilled water, and store in a refrigerator for not longer than 1 month. Further dilute with distilled water.
Dose levels: 40.0, 44.5, 50.0, 56.0, and 62.5 µg/ml
Incubation: 36° – 37.5°C

11.8   Demeclocycline

Test organism: *Staphylococcus aureus* ATCC 6538P
Maintain the culture, prepare the suspension, and store as described in Section 9.31.
Inoculum: Add 0.1 ml of the standardized suspension to each 100 ml of medium 3.
Standard: Dry the working standard as directed in Section 9.32. Prepare a 1 mg/ml stock solution in 0.1 *N* hydrochloric acid, and store in a refrigerator for not longer than 4 days. Further dilute with distilled water.
Dose levels: 0.080, 0.089, 0.100, 0.112, and 0.125 µg/ml
Incubation: 36° – 37.5°C

11.9   Dihydrostreptomycin

Test organism: *Klebsiella pneumoniae* ATCC 10031
Maintain the culture, prepare the suspension, and store as described in Section 9.31.
Inoculum: Add 0.1 ml of the standardized suspension to each 100 ml of medium 3.
Standard: Proceed as described in Section 9.32, except dry the working standard at 100°C for 4 hours. Prepare a 1 mg/ml stock solution in distilled water and store refrigerated for not longer than 1 month. Further dilute with distilled water.
Dose levels: 24.0, 26.8, 30.0, 33.5, and 37.5 µg/ml
Incubation: 36° – 37.5°C

11.10   Doxycycline

Test organism: *Staphylococcus aureus* ATCC 6538P
Maintain the culture, prepare the suspension, and store as described in Section 9.31.
Inoculum: Add 0.1 ml of the standardized suspension to each 100 ml of medium 3.
Standard: *Not dried.* Prepare a 1 mg/ml stock solution in 0.1 N hydrochloric acid, and store in a refrigerator for not longer than 5 days. Further dilute with distilled water.
Dose levels: 0.080, 0.089, 0.100, 0.112, and 0.125 μg/ml
Incubation: 36° – 37.5°C

11.11   Erythromycin

Test organism: *Klebsiella pneumoniae* ATCC 10031
Maintain the culture, prepare the suspension, and store as described in Section 9.31.
Inoculum: Add 0.1 ml of the standardized suspension to each 100 ml of medium 3.
Standard: Prepare a 10 mg erythromycin base per milliliter in methanol, then dilute to 1 mg/ml with distilled water. Further dilute with distilled water. Use the solutions the same day.
Dose levels: 0.48, 0.54, 0.60, 0.67, 0.75 μg/ml
Incubation: 36° – 37.5°C

11.12   Gramicidin

Test organism: *Streptococcus faecium* ATCC 10541
Maintain by subculturing regularly in 100 ml of medium 3. Inoculate 100 ml of medium 3 with a loopful of the stock culture, and incubate for 16 – 18 hours at 37°C. Store in a refrigerator for not longer than 24 hours.
Inoculum: Add 1.0 ml of the standardized suspension to each 100 ml of medium 3.
Standard: Dry the working standard as directed in Section 9.32. Prepare a 1 mg/ml stock solution in ethanol USP XX, and store in a refrigerator for not longer than 1 month. Further dilute with ethanol.
Dose levels: 0.032, 0.0356, 0.040, 0.0448, and 0.050 μg/ml
Incubation: 36° – 37.5°C

11.13   Kanamycin

Test organism: *Staphylococcus aureus* ATCC 6538P
Maintain the culture, prepare the suspension, and store as described in Section 9.31.
Inoculum: Add 0.2 ml of the standardized suspension to each 100 ml of medium 3.
Standard: *Not dried.* Prepare a 1 mg/ml stock solution in distilled water, and store in the refrigerator for not longer than 1 month. Further dilute with distilled water.
Dose levels: 8.0, 8.9, 10.0, 11.2, and 12.5 μg/ml
Incubation: 36° – 37.5°C

11.14 Lincomycin

Test organism: *Staphylococcus aureus* ATCC 6538P
Maintain the culture, prepare the suspension, and store as described in Section 9.31.
Inoculum: Add 0.1 ml of the standardized suspension to each 100 ml of medium 3.
Standard: *Not dried.* Prepare a 1 mg/ml stock solution in distilled water, and store in a refrigerator for not longer than 1 month. Further dilute with distilled water.
  Dose levels: 0.400, 0.447, 0.500, 0.559, and 0.625 μg/ml
  Incubation: 36°–37.5°C

11.15 Meclocycline

Test organism: *Staphylococcus aureus* ATCC 6538P
Maintain the culture, prepare the suspension, and store as described in Section 9.31.
Inoculum: Add 0.2 ml of the calibrated suspension to each 100 ml of medium 3.
Standard: *Not dried.* Prepare a 1 mg/ml solution in solution 13 and use the same day. Further dilute with distilled water.
  Dose levels: 0.048, 0.054, 0.060, 0.067, and 0.075 μg/ml
  Incubation: 36°–37.5°C

11.16 Methacycline

Test organism: *Staphylococcus aureus* ATCC 6538P
Maintain the culture, prepare the suspension, and store as described in Section 9.31.
Inoculum: Add 0.1 ml of the calibrated suspension to each 100 ml of medium 3.
Standard: Proceed as described in Section 9.32. Prepare a 1 mg/ml stock solution in distilled water, and store in a refrigerator for not longer than 7 days. Further dilute with distilled water.
  Dose levels: 0.048, 0.054, 0.060, 0.067, and 0.075 μg/ml
  Incubation: 36°–37.5°C

11.17 Minocycline

Test organism: *Staphylococcus aureus* ATCC 6538P
Maintain the culture, prepare the suspension, and store as described in Section 9.31.
Inoculum: Add 0.2 ml of the calibrated suspension to each 100 ml of medium 3.
Standard: *Not dried.* Prepare a 1 mg/ml stock solution in 0.1 N hydrochloric acid, and store for not longer than 2 days. Further dilute with distilled water.
  Dose levels: 0.068, 0.076, 0.085, 0.095, and 0.106 μg/ml
  Incubation: 36°–37.5°C

11.18   Neomycin

Test organism: *Klebsiella pneumoniae* ATCC 10031
Maintain the culture, prepare the suspension, and store as described in Section 9.31.
Inoculum: Add 0.2 ml of the standardized suspension to each 100 ml of medium 3.
Standard: Dry the working standard as described in Section 9.32. Prepare a 1 mg/ml of neomycin base stock solution in phosphate buffer pH 8 (solution 3), and store in a refrigerator for not longer than 1 month. Further dilute with distilled water.
Dose levels: 2.00, 2.24, 2.50, 2.80, and 3.12 μg/ml
Incubation: 36° – 37.5°C

11.19   Oxytetracycline

Test organism: *Staphylococcus aureus* ATCC 6538P
Maintain the culture, prepare the suspension, and store as described in Section 9.31.
Inoculum: Add 0.1 ml of the standardized suspension to each 100 ml of medium 3.
Standard: *Not dried.* Prepare a 1 mg/ml stock solution in 0.1 N hydrochloric acid, and store in a refrigerator for not longer than 4 days. Further dilute with distilled water.
Dose levels: 0.192, 0.215, 0.240, 0.268, and 0.300 μg/ml
Incubation: 36° – 37.5°C

11.20   Rifamycin Sodium

Test organism: *Escherichia coli* ATCC 10536
Maintain the culture, prepare the suspension, and store as described in Section 9.31.
Inoculum: Add 0.7 ml of the standardized suspension to each 100 ml of medium 3.
Standard: Prepare a 1 mg/ml stock solution in distilled water, and store in a refrigerator for not longer than 1 day. Further dilute with distilled water.
Dose levels: 0.80, 0.89, 1.00, 1.12, and 1.25 μg/ml
Incubation: 36° – 37.5°C

11.21   Rolitetracycline

Test organism: *Staphylococcus aureus* ATCC 6538P
Maintain the culture, prepare the suspension, and store as described in Section 9.31.
Inoculum: Add 0.1 ml of the standardized suspension to each 100 ml of medium 3.
Standard: Dry the working standard as described in Section 9.32. Prepare a 1 mg/ml stock solution in distilled water, and store in a refrigerator for not longer than 1 day. Further dilute with distilled water.
Dose levels: 0.192, 0.215, 0.240, 0.268, and 0.300 μg/ml
Incubation: 36° – 37.5°C

11.22 Spectinomycin

Test organism: *Escherichia coli* ATCC 10536
Maintain the culture, prepare the suspension, and store as described in Section 9.31.
Inoculum: Add 0.1 ml of the standardized suspension to each 100 ml of medium 3.
Standard: Not dried. Prepare a 1 mg/ml stock solution in distilled water, and store in a refrigerator for not longer than 1 month. Further dilute with distilled water.
Dose levels: 24.0, 26.8, 30.0, 33.5, and 37.5 μg/ml
Incubation: 36°–37.5°C

11.23 Streptomycin

Test organism: *Klebsiella pneumoniae* ATCC 10031
Maintain the culture, prepare the suspension, and store as described in Section 9.31.
Inoculum: Add 0.1 ml of the standardized suspension to each 100 ml of medium 3.
Standard: Dry the working standard as described in Section 9.32. Prepare a 1 mg/ml stock solution in distilled water, and store in a refrigerator for not longer than 1 month. Further dilute with distilled water.
Dose levels: 24.0, 26.8, 30.0, 33.5, and 37.5 μg/ml
Incubation: 36°–37.5°C

11.24 Tetracycline

Test organism: *Staphylococcus aureus* ATCC 6538P
Maintain the culture, prepare the suspension, and store as described in Section 9.31.
Inoculum: Add 0.1 ml of the standardized suspension to each 100 ml of medium 3.
Standard: Not dried. Prepare a 1 mg/ml stock solution in 0.1 N hydrochloric acid, and store for not longer than 1 day. Further dilute with distilled water.
Dose levels: 0.192, 0.215, 0.240, 0.268, and 0.300 μg/ml
Incubation: 36°–37.5°C

11.25 Tobramycin

Test organism: *Staphylococcus aureus* ATCC 6538P
Maintain the culture, prepare the suspension, and store as described in Section 9.31.
Inoculum: Add 0.15 ml of the standardized suspension to each 100 ml of medium 3.
Standard: Not dried. Prepare a 1 mg/ml stock solution in distilled water, and store in a refrigerator for not longer than 2 weeks. Further dilute with distilled water.
Dose levels: 2.000, 2.236, 2.500, 2.795, and 3.125 μg/ml
Incubation: 36°–37.5°C

11.26   Troleandomycin

Test organism: *Klebsiella pneumoniae* ATCC 10031
Maintain the culture, prepare the suspension, and store as described in Section 9.31.
Inoculum: Add 0.1 ml of the standardized suspension to each 100 ml of medium 3.
Standard: Dry the working standard as described in Section 9.32. Prepare a 1 mg/ml stock solution in 80% aqueous 2-propanol (solution 15), and use the same day. Further dilute with distilled water.
Dose levels: 20.00, 22.25, 25.00, 28.00, and 31.25 μg/ml
Incubation: 36° – 37.5°C

11.27   Tyrothricin

Test organism: *Streptococcus faecium* ATCC 10541
Maintain the culture, prepare the suspension, and store as described in Section 11.12.
Inoculum: Add 1.0 ml of the standardized suspension to each 100 ml of medium 3.
Standard: Use the gramicidin working standard as described in Section 11.12.
Dose levels: As described for gramicidin in Section 11.12
Incubation: 36° – 37.5°C

11B   VITAMINS

11.28   Introduction

The theoretical and practical considerations relating to the tube assay of vitamins were fully discussed in Chapter 6. In the following pages some examples will be described for the tube assay of the more important vitamins in pharmaceutical products. The details here given must be supplemented by relevant information in preceding chapters for completeness.

The general principles of culture maintenance regarding the lactobacilli were introduced in Chapter 2, but from a practical point of view, especially in a busy assay laboratory, it is equally important to ensure that the cultures are readily available at short notice. It is still important to passage the cultures alternately on solid and liquid media, as explained in Chapter 2, but instead of allowing a lapse of 1–2 weeks between subcultures and therefore between changes from solid to liquid medium and vice versa, run the two cultures concurrently and cross over the two cultures at each subculturing. Details of this crossover technique will be described more fully in the following pages.

The vitamin depletion method of inoculum preparation introduced in Chapter 10 for the diffusion assays of vitamins can be just as successfully applied for the tube assays of vitamins. Several examples will be found in the following pages offering fully tested methods which worked well in the hands of the authors. The more traditional inoculum preparation, involving centrifugation followed by repeated washing of the bacterial suspension, is also described in some of the examples which follow. This same method can also be used for all the other test organisms, especially if an urgent assay is required and time is short, or if there is a centrifuge normally available.

The next section will describe general techniques applicable to all the tube assays followed by examples of methods for some of the commonly occurring vitamins.

11.29 Common Techniques

11.29.1 Preparation of Equipment

Test tubes: Use rimless heat-resistant test tubes of suitable size, typically 16×150 mm. Ordinary washing of test tubes can leave behind traces of fatty acids or detergents which can interfere with the growth of an assay organism, producing either suppression or enhancement of growth, resulting in spurious turbidities. To avoid this effect, clean all test tubes as follows:

Soak the test tubes overnight in a 2% (v/v) aqueous solution of Decon 90 concentrate (see Appendix 1). Discard the solution from the tubes and rinse them immediately with plenty of tap water, because the detergent must not be allowed to dry on the glass surface. Give a final rinse of distilled water, and dry the tubes upside down stacked in suitable wire baskets in an incubator or glassware-drying cabinet.

Before the tubes are used for an assay, fill sufficient tubes, which are stacked upright in a wire basket, with freshly distilled water, cover loosely with aluminum foil, and autoclave at 121 °C for 15 minutes. Allow to cool until they are safe to handle, then carefully drain the tubes by transferring them upside down into another wire basket. Cover with aluminum foil and dry overnight in an incubator. The following morning, label the basket "ready for tube assays." See also Section 7.14 for further details.

Aluminum caps: Use suitable caps with straight sides which slide over the tubes easily. (Oxoid caps are suitable.) Although the caps should not come in contact with the assay medium, they also have to be thoroughly cleaned. Soak the caps in a very dilute solution of Decon 90 (e.g., one test tube capful to a sinkful of cold water) for about 1 hour. Rinse well first with plenty of tap water followed by distilled water, then drain and dry in an incubator.

Test tube racks: It is recommended to use metal racks which can accommodate 36 tubes in 12 columns of 3 each. Mark each rack with the appropriate design number (see Appendix 8), having first written the correct code number over each hole as shown in Fig. 11.1. Three separate racks will be needed for each assay, as indicated by the example in Fig. 11.1. The numbers 1–36 represent a complete assay of up to six samples with a set of standard solutions, each one in triplicate.

Pasteur pipettes: These are washed, rinsed in distilled water, dried, plugged with cotton wool, individually wrapped in aluminum foil, and sterilized in a hot-air oven.

Graduated pipettes: Wash and rinse pipettes in tap water, give a final rinse in distilled water, and dry in an incubator. Plug with cotton wool, pack into a suitable metal canister, and sterilize by autoclaving at 121 °C for a minimum of 20 minutes.

Volumetric pipettes: After washing with tap water, rinse with distilled water and allow to dry in an incubator. After use put all pipettes into a tall pipette jar containing dilute Decon 90 solution.

Beakers and other equipment: Beakers and any other equipment such as glass rods, spatulas, or anything else used in preparation must be scrupulously cleaned to ensure that all traces of vitamins have been fully removed.

Rack 1. Design 3.

12	10	17	25	35	16	3	36	24	1	28	11
0	0	0	0	0	0	0	0	0	0	0	0
14	15	27	21	29	6	33	2	18	31	9	23
0	0	0	0	0	0	0	0	0	0	0	0
19	26	32	13	20	4	8	34	30	22	5	7
0	0	0	0	0	0	0	0	0	0	0	0

Rack 2. Design 18.

19	10	12	6	32	22	24	14	11	28	9	35
0	0	0	0	0	0	0	0	0	0	0	0
25	33	21	1	20	4	18	36	2	13	15	23
0	0	0	0	0	0	0	0	0	0	0	0
27	8	5	31	29	7	16	3	34	26	17	30
0	0	0	0	0	0	0	0	0	0	0	0

Rack 3. Design 31.

2	25	16	11	29	13	20	10	34	6	12	32
0	0	0	0	0	0	0	0	0	0	0	0
23	27	5	1	33	3	36	24	18	8	22	31
0	0	0	0	0	0	0	0	0	0	0	0
26	14	19	7	28	30	9	15	4	17	35	21
0	0	0	0	0	0	0	0	0	0	0	0

Fig. 11.1. Typical random arrangement of standard and sample solutions in test tubes using three separate 3 × 12 test tube racks for turbidimetric assays. (Other suitable designs can be found in Appendix 8.) Positions 1–5 are taken up by standard solutions in ascending order of concentration and the other positions by sample dilutions in the same order. Note that each solution is replicated three times but appears once only in each rack. Position 36 is reserved for the inoculated blank. (See Table 11.2 for more details.)

11.29.2 Methodology

Assay media: It is more cost effective to use readily available quality-controlled dehydrated complete-formula vitamin assay media, which are obtainable commercially from reputable manufacturers (e.g., Difco, BBL). The authors have always relied on Difco media. Always store the dehydrated media in a refrigerator and follow the manufacturer's instruction for preparation. Allow approximately 100 ml of double-strength medium for each sample or standard.

Assay method: Robinson (1971) had shown that there was a significant statistical rack-to-rack bias when standard and sample tubes were set up in separate racks, which used to be common practice. He constructed random designs to eliminate this bias, and examples of these designs can be found in Appendix 8. Each rack carries its own design, as shown in Fig. 11.1, and any three racks taken at random can be used to accommodate one standard and up to six samples, each at five dose levels and each dose level in triplicate. The racks need not be filled completely, and it is possible to assay only one sample against one standard; however, it is important to place the tubes in the racks following the appropriate design and simply leave the unused positions empty.

Number 1 refers to the lowest dose of the standard, number 2 to the next dose of the standard, and so on for all doses of the standard. The following numbers refer to the unknowns in the same order and finally, number 36 is used for the inoculated blank. Table 11.2 gives full details of the arrangement favored by the authors. The actual practical steps are as follows.

Dilute the standard and samples to the assay levels indicated under the individual methods and, using a 5-ml graduated pipette, transfer 1 ml of the standard followed by 4 ml of distilled water to three separate tubes, and place the three tubes into their respective positions marked by number 1 in each of the three racks. Add 2 ml of standard and 3 ml of distilled water to the next three tubes and place them in the racks under number 2. Continue with triplicate tubes for each tube number as indicated in Table 11.2, and place the tubes in the racks as shown by the example in Fig. 11.1.

Preferably using an automatic pipetting syringe, add 5 ml of the appropriate double-strength medium to each of the tubes in column order (e.g., in Fig. 11.1 add medium to the tubes in positions 12, 14, and 19, then to 10, 15, and 26). Continue in column order even if some of the positions are not occupied. Cover each tube with an aluminum cap and sterilize by heating at 105 °C for 20 minutes. After cooling to room temperature as rapidly as possible, inoculate the tubes in column order and shake the rackful of tubes carefully to mix. Incubate the tubes as directed under the individual methods.

Standards: Most vitamins are available as pure chemical compounds. Reference standards can be obtained from various sources such as the USP, FDA, or WHO, but for most practical purposes it is quite satisfactory to select the chemically purest production supply and reserve a small quantity—say, 5–100 g—depending on the expected usage. Keep the standard in a tightly closed container in a refrigerator.

Sample treatment: The majority of pharmaceutical preparations, such as multivitamin tablets and syrups, contain added pure vitamins and in many cases will require only straightforward dilution with water. Naturally occurring substances like milk, cereals, yeast extract, and serum will need to be extracted by one of the published methods, depending on the nature of the product. Consult the literature for specialized requirements, and the most suitable treatment as the need arises. Guidelines are given by György (1950), Barton-Wright (1952), and Kavanagh (1963 and 1972).

Reading opacities: After incubation stop the growth by adding one drop of formaldehyde solution to each tube in column order, and leave to stand at room temperature for 10 minutes. Then read the turbidities in column order. Vortex each tube briefly before reading. Record the

Table 11.2

Five-Dose Level Tube Assays with Standard and Up to Six
Samples

Tube number	Treatment solution	Sample volume (ml)	Distilled water (ml)	Double-strength medium (ml)
1	Standard	1	4	5
2	Standard	2	3	5
3	Standard	3	2	5
4	Standard	4	1	5
5	Standard	5	0	5
6	Sample 1	1	4	5
7	Sample 1	2	3	5
8	Sample 1	3	2	5
9	Sample 1	4	1	5
10	Sample 1	5	0	5
11	Sample 2	1	4	5
12	Sample 2	2	3	5
13	Sample 2	3	2	5
14	Sample 2	4	1	5
15	Sample 2	5	0	5
16	Sample 3	1	4	5
17	Sample 3	2	3	5
18	Sample 3	3	2	5
19	Sample 3	4	1	5
20	Sample 3	5	0	5
21	Sample 4	1	4	5
22	Sample 4	2	3	5
23	Sample 4	3	2	5
24	Sample 4	4	1	5
25	Sample 4	5	0	5
26	Sample 5	1	4	5
27	Sample 5	2	3	5
28	Sample 5	3	2	5
29	Sample 5	4	1	5
30	Sample 5	5	0	5
31	Sample 6	1	4	5
32	Sample 6	2	3	5
33	Sample 6	3	2	5
34	Sample 6	4	1	5
35	Sample 6	5	0	5
36	Blank	0	5	5

responses as they are read one after the other for one rack at a time, and unscramble the responses only after all of the tubes have been read. Refer to Chapter 6 for further details or to Hewitt (1977) for full treatment of this subject.

Calculation of results: A full account of suitable calculations is given by Hewitt (1977), with worked examples and full explanations. However, it is often useful to carry out a quick initial check on the results by constructing graphs of both standard and samples. A great deal of useful information can be obtained from these graphs after a simple inspection. For example, if the standard and sample lines cross, then the assay is invalid. The position of the sample curve in relation to the standard curve will indicate whether the potency of the sample is higher or lower than that of the standard, and so on.

To obtain an estimate of sample potency, proceed as follows:

(1) Sum triplicate readings for each dose level.
(2) Draw the standard curve after plotting on a graph paper the standard reading against each dose level from 1 to 5 ml.
(3) Similarly plot the sample dose readings.
(4) Read off potencies for each dose level of the sample directly from the standard curve.
(5) Divide each graph reading of the sample by its dose level.
(6) Average all the figures which are in good agreement.
(7) Calculate the result of the sample using the formula:

Average reading at 1 ml × assay level × factor of standard × dilution of sample = ... µg/ml of vitamin in unknown sample

11.30 Thiamine

Test organism: Lactobacillus fermentum NCIB 6991, ATC 9338
Maintain the organism by at least monthly subculture alternately on solid and liquid culture media. For best results, inoculate the two cultures concurrently: on each occasion inoculate from the microassay culture agar stab growth into fresh microinoculum broth and from the microinoculum broth culture into a fresh microassay culture agar stab. Incubate both cultures at 37°C for 48 hours. Cross-inoculate from broth to stab and from stab to broth a day or two before preparation of the inoculum. Store the cultures between subculturing in the refrigerator.

Inoculum: Aseptically transfer 0.25 ml of a uniformly suspended recent broth culture into 10 ml of sterile microinoculum broth 3 days before the assay is planned, and incubate overnight at 37°C. Shake to suspend the growth and transfer the whole overnight culture to a Roux bottle (or a 500-ml medical flat bottle) containing about 80 ml of sterile *single-strength* thiamine assay medium. Incubate the bottle (lying on its side so that the medium occupies maximum surface area) at 37°C overnight. The following day transfer 20 ml of the suspended growth into another large bottle containing about 80 ml of sterile *single-strength* thiamine assay medium, and again incubate lying on its side at 37°C overnight. When the assay tubes are ready to be inoculated, transfer a sufficient amount of the suspended overnight culture dropwise into another bottle of *single-strength* thiamine assay medium to produce just-visible opacity (~90% light transmission). The inoculum is now ready to use.

Assay medium: Use Difco thiamine assay medium or equivalent (code 0326). Measure the required volume of distilled water into a large beaker, cover loosely with aluminum foil, and bring the water to the boil. Weigh an amount of the dehydrated medium according to the manufacturer's instructions to produce double-strength liquid medium, typically 8.5 g for each 100 ml of distilled water. When the water starts to boil, remove the heat source and slowly add the medium while stirring continuously. Take care because it is quick to boil over. Reheat, and as soon as it starts to boil again, carefully remove from the heat and allow to cool in the dark.

Standard solutions: Prepare a stock solution of the working standard as described in Section 10.2, and further dilute with distilled water to the assay concentration of 0.02 μg/ml.

Sample solutions: Dilute the samples with distilled water to the assay concentration of 0.02 μg/ml. Calculate the approximate dilution required by using the following equation:

$$\text{Dilution of sample} = \frac{\text{assumed potency of sample (μg/ml)}}{\text{assay concentration (μg/ml)}} \qquad (11.1)$$

For example, if assumed potency of sample is 0.3 μg/ml, then

$$\frac{0.30}{0.02} = 15$$

therefore dilute sample 1 : 15, typically 10 ml → 150 ml.

Note: Protect standard and sample solutions from light.

Assay procedure: Fill the tubes as described in Section 11.29 and, after autoclaving, cool the tubes rapidly and inoculate them in column order with one drop of the standardized inoculum per tube from a Pasteur pipette. Shake each rackful of tubes carefully to mix the contents, and incubate at 37°C for approximately 16 hours.

Measurement of responses and calculations: Proceed as described in Section 11.29.

11.31   Riboflavin

Test organism: *Lactobacillus casei* subsp. *rhamnosus* NCIB 8010, ATCC 7469
Maintain the organism as described in Section 11.30.

Inoculum: Proceed as described in Section 11.30, except use *single-strength* Difco riboflavin assay medium or equivalent in place of the single-strength thiamine assay medium.

Assay medium: Use Difco riboflavin assay medium (code 0325) or equivalent. Prepare the medium as described in Section 11.30, except use 4.8 g of dehydrated assay medium for each 100 ml of distilled water.

Standard solutions: Prepare a stock solution of the working standard as described in Section 10.3, and further dilute with distilled water to the assay concentration of 0.02 μg/ml.

Sample solutions: Dilute the samples with distilled water to the assay concentration of 0.02 μg/ml. Calculate the approximate dilution required by using Eq. (11.1). For example, if the assumed potency of the sample is 0.6 μg/ml, then 0.60/0.02 = 30; therefore, dilute sample 1 : 30, typically 10 ml → 300 ml.

Note: Protect standard and sample solutions from light.

Assay procedure: Proceed as described in Section 11.30 and incubate the tubes at 37°C for 16 – 18 hours.

Measurement of responses and calculations: Proceed as described in Section 11.29.

Note: Riboflavin 5-phosphate can also be assayed by the same method against the appropriate standard, but riboflavin 5-phosphate is very sensitive to light and solutions must be protected from light throughout dilutions and assay.

11.32   Pyridoxine

Test organism: *Saccharomyces uvarum (carlsbergensis)* NCTC 4228, ATCC 9080

Maintain the organism by monthly subculture on Sabouraud's agar slants and incubation at $27° - 30°C$ for approximately 48 hours. Store in the refrigerator after incubation. Two days before the culture is needed for the assay, subculture the organism on a Sabouraud's agar slant and incubate at $27° - 30°C$ for about 48 hours.

Inoculum: While the assay tubes are cooling (see below), mix a loopful of the fresh slant culture with sterile distilled water to just visible turbidity ($\sim 90\%$ light transmission). Allow at least 1 ml of distilled water for each assay tube.

Assay medium: Use Difco pyridoxine Y medium (code 0951) or equivalent. Prepare the medium as described in Section 11.30, except use 5.3 g of dehydrated assay medium for each 100 ml of distilled water.

Standard solutions: Prepare a stock solution of the working standard as described in Section 10.4, and further dilute with distilled water to the assay concentration of 0.004 μg/ml.

Sample solutions: Dilute the samples with distilled water to the assay concentration of 0.004 μg/ml. Calculate the approximate dilution required by using Eq. (11.1). For example, if the assumed potency of the sample is 0.2 μg/ml, then $0.2/0.004 = 50$; therefore, dilute sample 1 : 50, that is, typically 5 ml $\rightarrow$ 250 ml.

Assay procedure: After sterilization of the assay tubes, cool them to $23° - 25°C$ and inoculate each tube in column order using exactly 1.0 ml of the inoculum. If a sterile 5-ml graduated pipette is used, five assay tubes can be inoculated with a pipetteful of inoculum. Shake the rackful of tubes carefully to mix the contents, and incubate at $27° - 30°C$ for approximately 22 hours.

Measurement of responses and calculations: Proceed as described in Section 11.29.

11.33   Cyanocobalamin

Test organism: *Lactobacillus delbrueckii* subsp. *lactis (leichmanii)* NCTC 8118, ATCC 7830

Maintain the organism as described in Section 11.30.

Inoculum: Start the inoculum a day before the assay is planned by transferring a loopful of the latest stab culture into Difco microinoculum broth (code 0320) or equivalent, and incubate at 37°C overnight. While the assay tubes are being autoclaved, remove the overnight culture from the incubator, shake to suspend the growth and spin down in a centrifuge for 5 minutes at top speed, or until a firm deposit has formed. Carefully decant the clear supernatant broth, resuspend in sufficient sterile saline, and spin down again. Decant the clear supernatant liquid, resuspend in more sterile saline, and spin down twice more as before. After the final spin, resuspend the deposited cells in saline and add sufficient of the suspension to about 100 ml of sterile saline to produce just visible opacity ($\sim 90\%$ light transmission). The inoculum is now ready to use.

Note: Do not allow the compacted culture to stand without saline.

Assay medium: Use Difco B_{12} assay medium USP (code 0457) or equivalent. Prepare the medium as described in Section 11.30, except use 8.5 g of the dehydrated B_{12} assay medium for each 100 ml of distilled water.

Standard solutions: Prepare a stock solution of the working standard as described in Section 10.5, and further dilute with distilled water to the assay concentration of 0.00005 μg/ml (0.05 ng/ml).

Sample solutions: Dilute the samples with distilled water to the assay concentration of 0.00005 μg/ml. Calculate the approximate dilution required by using Eq. (11.1). For example, if the assumed potency of the sample is 0.1 μg/ml, then 0.1/0.00005 = 200; therefore, dilute sample 1 : 200, that is, typically 5 ml → 50 ml : 5 ml → 100 ml.

Assay procedure: After sterilization of the assay tubes, cool them to 23°–25°C and inoculate each tube in column order with one drop of the inoculum using a sterile Pasteur pipette. Shake each rackful of tubes carefully to mix the contents, and incubate at 37°C for approximately 16 hours.

Measurement of responses and calculations: Proceed as described in Section 11.29.

Note: This method can be used also for the assay of hydroxocobalamin, provided the appropriate standard is prepared and the addition of potassium cyanide is omitted. Protect the standard and sample solutions from light throughout their preparation as well as the assay tubes during heating and cooling. Hydroxocobalamin solutions require semidarkness during handling.

11.34   Folic Acid

Test organism: *Lactobacillus casei* subsp. *rhamnosus* NCIB 8010, ATCC 7469
Maintain the organism as described in Section 11.30.

Inoculum: One day before the assay is due, take the most recent broth culture from the refrigerator and, after thorough mixing, transfer about 2.5 ml into 10 ml of fresh microinoculum broth, and incubate at 37°C overnight. While the assay tubes are being autoclaved, remove the culture from the incubator and spin down the culture in a centrifuge as detailed in Section 11.33.

Assay medium: Use Difco folic acid casei medium (code 0822) or equivalent. Prepare the medium as described in Section 10.30, except use 9.4 g of the dehydrated folic acid casei medium for each 100 ml of distilled water.

Standard solutions: Prepare a stock solution of the working standard as described in Section 10.7, and further dilute with distilled water to the assay concentration of 0.0002 μg/ml.

Sample solutions: Dilute samples with distilled water to the assay concentration of 0.0002 μg/ml. Calculate the approximate dilution required by using Eq. (11.1). For example, if the assumed potency of the sample is 0.010 μg/ml, then 0.010/0.0002 = 50; therefore, dilute sample 1 : 50, that is, typically 5 ml → 250 ml.

Assay procedure: After sterilization of the assay tubes, cool them to 23°–25°C and inoculate each tube in column order with one drop of the inoculum using a sterile Pasteur pipette. Shake each rackful of tubes carefully to mix the contents, and incubate at 37°C for approximately 16 hours.

Measurement of responses and calculations: Proceed as described in Section 11.29.

Note: Cotton wool plugs could contain traces of folic acid and must not be used either for the cultures or for the assay tubes.

11.35   Niacin

Test organism: *Lactobacillus plantarum* NCIB 8030, ATCC 8014
Maintain the organism as described in Section 11.30.

Inoculum: Proceed as described in Section 11.30, except use *single-strength* Difco niacin assay medium (code 0322) or equivalent instead of single-strength thiamine assay medium.

Assay medium: Use Difco niacin assay medium (code 0322) or equivalent. Prepare the medium as described in Section 11.30, except use 7.5 g of the dehydrated niacin medium for each 100 ml of distilled water.

Standard solutions: Prepare a stock solution of the working standard as described in Section 10.8, and further dilute with distilled water to the assay concentration of 0.02 μg/ml.

Sample solutions: Dilute samples with distilled water to the assay concentration of 0.02 μg/ml. Calculate the approximate dilution required by using Eq. (11.1). For example, if the assumed potency of the sample is 0.30 μg/ml, then 0.30/0.02 = 15; therefore, dilute sample 1 : 15, that is, typically 10 ml → 150 ml.

Assay procedure: After sterilization of the assay tubes, cool them to 23°–25°C and inoculate each tube in column order with one drop of the inoculum using a sterile Pasteur pipette. Shake each rackful of tubes carefully to mix the contents, and incubate at 37°C for approximately 16 hours.

Measurement of responses and calculations: Proceed as described in Section 11.29. See also example in Chapter 6.

Note: Niacin tube assays can be recommended to anyone trying to carry out tube assay of vitamins for the first time.

11.36 Pantothenate

Test organism: *Lactobacillus plantarum* NCIB 8030, ATCC 8014
Maintain the organism as described in Section 11.30.

Inoculum: Transfer 0.25 ml from the most recent broth culture into 10 ml sterile microin-oculum broth *2 days* before the assay is to be carried out, and incubate overnight at 37°C. Transfer the whole overnight culture to a Roux bottle (or to a 500-ml medical flat bottle fitted with a screw cap) containing about 80 ml of *single-strength* Difco pantothenate AOAC medium (code 0816) or equivalent. Lay the bottle flat on its side, and incubate overnight at 37°C. When the assay tubes are ready to be inoculated, transfer dropwise a sufficient amount of the suspended overnight culture into another bottle of single-strength pantothenate medium to produce just-visible opacity (~90% light transmission). The inoculum is now ready to use.

Assay medium: Use Difco pantothenate assay medium (code 0816) or equivalent. Prepare the double-strength assay medium as described in Section 11.30, except use 7.3 g of the dehydrated pantothenate AOAC medium for each 100 ml of distilled water.

Standard solutions: Prepare a stock solution of the working standard as described in Section 10.9, and further dilute with distilled water to the assay concentration of 0.02 μg/ml.

Sample solutions: Dilute the samples with distilled water to the assay concentration of 0.02 μg/ml. Calculate the approximate dilution required by using Eq. (11.1). For example, if the assumed potency of the sample is 0.6 μg/ml, then 0.6/0.02 = 30; therefore, dilute sample 1 : 30, that is, typically 10 ml → 300 ml.

Assay procedure: After sterilization, cool the assay tubes rapidly to 23°–25°C and inoculate each tube in column order with one drop of the inoculum using a sterile Pasteur pipette. Shake each rackful of tubes carefully to mix the contents and incubate at 37°C for approximately 16 hours.

Measurement of responses and calculations: Proceed as described in Section 11.29.

References

Barton-Wright, E. C. (1952). "The Microbiological Assay of the Vitamin B Complex and Amino Acids." Pitman, London.

György, P., ed. (1950). "Vitamin Methods," Vol. 1. Academic Press, New York.

Hewitt, W. (1977). "Microbiological Assay." Academic Press, New York.

Kavanagh, F. W. (1963 and 1972). "Analytical Microbiology," Vols I and II. Academic Press, New York and London.

Kuzel, N. R., and Kavanagh, F. W. (1971). *J. Pharm. Sci.* **60,** 764 and 767.

Robinson, W. D. (1971). Personal communication.

CHAPTER 12

ASSAY REPLICATION

12.1 Introduction

The principles of replicate assays to achieve greater precision and accuracy have been clearly explained by Hewitt (1977), who has also given guidance on the evaluation of replicate results.

It is a well-known fact that the generation of precise and accurate results demands a great deal of effort on the part of the analytical microbiologist, and careful attention to detail, as outlined in previous chapters, is essential. Of course, one has to be careful that the results are meaningful and genuine. A distinguished chief analyst used to say that if two results agreed, it only proved that they could be both wrong. The authors have certainly seen examples when close agreement of assay results was no guarantee that they were correct, because investigation into the assay method often revealed a bias in the results. The bias can be due to setting the dose levels in an insensitive range, to faulty designs or practices, and can go undetected for long periods.

In the following pages it will be assumed that all the assay results have been generated in accordance with the principles and techniques described in this book and by the conscious elimination of all avoidable bias.

12.2 Replication for Routine Samples

Each routine sample should be assayed in duplicate, preferably by two separate and independent assistants, who should each weigh their own sample and prepare their own dilutions. The standard solutions could be prepared by only one of the assistants. If there is only one person available, the duplication of assay results either on the same day or preferably on another day would be quite acceptable.

The mean of these two results should then be reported, provided that the two results were in fairly good agreement. Previously published guidelines, such as the so-called range/mean test introduced by Lees and Tootill (1955) and quoted by Hewitt (1977), and the 22% range of two results relative to their mean quoted by Simpson (1963), have been overtaken by continual progress. Better understanding of the microbiological assay system, improved techniques, improved equipment, and the introduction of automation have all helped to produce more consistent results than were possible a few years ago.

If the instructions described in this book are followed closely, beginners should be able to achieve duplicate figures which are within $\pm 10\%$ of the mean result. With increasing experience these limits should be gradually reduced to within $\pm 5\%$ of the mean.

If these arbitrary targets are not realized on the first occasion, then the assays should be repeated, again in duplicate, until the suggested targets are achieved. In the writer's (Vincent) laboratory duplicate results consistently fall well below $\pm 2.5\%$ of the mean, using single-sample two-dose level 8×8 Latin square assays and experienced assistants.

The mean of duplicate results is normally quite satisfactory as a production control test when the results are required to come within predefined limits or when they have to conform to a label claim.

12.3   Replication for Stability Samples

When the analyst is examining stability samples it is more important to obtain a really firm estimate of the potency at the various "takeoff" points and storage temperatures, because often these results will determine what life to expiry can be assigned to a product. It may be required to establish, for example, whether or not the active ingredient has lost 2 or 3% of its activity under a specified storage condition. In these cases it is not uncommon to reassay each sample on three or four separate occasions in duplicate and then to combine all results to produce an arithmetic mean.

A similar approach is needed when it is necessary to validate that an analytical procedure, such as a high-performance liquid chromatography (HPLC) method is a stability-indicating assay. The author's (Vincent) laboratory was involved in some of these validation assays using single-sample 8×8 Latin square designs on large plates, and the results were reported by Coomber *et al.* (1982), for example.

12.4   Replication for Raw Materials

Accurate potency determination of an antibiotic drug substance is extremely important because, whether the company is selling or buying, the price is more often than not fixed to potency per milligram of the active ingredient. Obviously, if the potency is underrated by the manufacturer's assay laboratory this could mean a financial loss to the company, and if the buyer's laboratory or their independent outside assessor correctly arrives at a higher figure then the buying company is able to gain some extra profits. The contract between the two companies may have an agreement which specifies payment on a potency calculated as the mean assay figure of the results from the two assay laboratories.

An interesting situation arose some time ago between the author's (Vincent) and another manufacturer's laboratories. It was noticed by the purchasing department that the protocol accompanying each batch of raw material always quoted significantly higher potency than was possible to obtain in the buyer's laboratory. When the reason for the discrepancy was investigated by the author, it was discovered that the two laboratories used different working standards with widely differing assigned potencies. The working standard used by the suppliers had an inflated high potency which produced the upward bias in the results. After an agreement amicably arrived at between the two laboratories to use the same working standard in both laboratories in the future, the previously observed discrepancy between the two sets of results disappeared.

Another reason why accurate potency assessment is important for raw drug substances is that when they are incorporated into formulated products, such as tablets, creams, and ointments, the manufacturer has to add just the right amount to make sure that the label claim is met throughout the life of the product. The amount added must take into account the overage and degradation of the drug substance; at the same time it must be ensured that none of the material is used wastefully.

In all these cases it is usual to carry out what is known as "5-day assays." This means duplicate assays on a minimum of 5 separate days, that is, not less than 10 individual assay results. With this many replicate assay figures, it is well worth working out the best estimate of the mean result together with confidence limits. Hewitt (1977) gives suitable examples of calculations.

12.5 Replication for Calibration of In-House Standards

It should be obvious from what has been said so far about the necessity to replicate assays for various purposes that setting a reliable potency figure on a house standard cannot be achieved in a single working day. It must be remembered that every unknown sample is compared to the standard in biological assays and if the standard is wrong then every single sample potency will also be wrong.

National or international standards are available for the sole purpose of setting up house standards, and on no account should they be used for routine daily assays. Three-dose level assays should be used to allow checking for linearity of responses.

The main steps involved in setting up a house standard can be summarized as follows:

(1)   Obtain an up-to-date reference standard.

(2)   Carry out some mock assays to validate the assay system before the reference standard is opened.

(3)   Carry out replicate assays of the proposed house standard against the reference standard.

(4)   Combine separate results into a mean potency and calculate confidence limits.

(5)   Assign potency to assay working standard, label, and store.

Reference standards are obtainable from various sources, such as the International Laboratory for Biological Standards (National Institute for Biological Standardisation and Control, Blanche Lane, South Mimms, Potters Bar, EN6 3QG England) and from other national or pharmacopoeial authorities. A comprehensive list with full addresses is given in Appendix 5.

The reference standards are supplied only in very limited quantities; therefore, every effort should be made not to waste them in any way but to obtain maximum results for a minimum of effort. Storage instructions received with the reference standard must be observed, especially if some delay is likely before the actual comparative assays take place.

It is a sensible precaution to make absolutely sure that the selected assay system is working properly before the actual calibration of the working standard is undertaken. There is usually a plentiful supply of the proposed new working standard, and this should be used to validate the assay conditions. Once validated, nothing should be changed; that is, use the same batch of assay medium, the same batch of spore suspension, the same dose levels, the same incubation temperature, and so on throughout the calibration assays.

If possible, two highly trained and experienced assistants should be available for this type of assay. In any case, the assistants should show an interest in assay work and they should have a definite aptitude for the work. (Consult Chapter 7 for training.)

Organizing the calibration assays will depend on how many reference standards could be obtained. If two reference standards were received, then it is safer not to use both reference standards at the same time but to reserve one of them just in case anything went wrong with the assay.

If only a single reference standard was obtained, which is the most likely situation, then the following procedure is recommended.

On day 1, assistant A should weigh an appropriate amount of the reference standard (usually about half the contents) using a four- or five-place analytical balance, and the rest of the standard should be adequately sealed and returned into storage, usually in the refrigerator. The same assistant should then prepare the three dose levels of the reference standard, and the same person should also weigh two separate aliquots of the proposed new standard. Three-dose level solutions of both of these aliquots should then be prepared by the same assistant, and finally, assistant A should set up an assay of the two separate weighings of the proposed working standard against the reference standard. Single-sample large plates, such as 8×8, 12×12, or 6×6 Latin square designs should be used in preference to petri dishes.

On the same day, assistant B should also weigh two separate amounts of the proposed new standard, prepare the appropriate three-dose level solutions, and assay these two sets against the reference standard solutions prepared by assistant A. This plan will produce a total of four results for the first day.

On day 2, reverse the process so that assistant B weighs and prepares solutions of the reference standard in addition to the two weighings and three-dose level solutions of the proposed new standard, while assistant A will prepare only the two weighings of the proposed new standard, thus producing another four results.

If it is known that the standard stock solution is stable at refrigeration temperature, as most antibiotics are, the system outlined above can be repeated on each day following preparation of the reference standard solutions. After making fresh dilutions of the stock solution, these can be used to assay fresh weighings of the proposed working standard, this way doubling the number of results from 8 to 16. Figure 12.1 is a flow diagram of these calibration assays (without refrigeration of the reference standard stock solution).

As stated earlier, if two reference standards were obtained, only one of them should be used on the first day. In this case let both assistants weigh independently from the same ampul of reference standard (usually about half the contents), and let each of them produce two weighings of the proposed new standard. Both assistants should then prepare all their own three-dose level solutions and assay the two samples of proposed standard against their own reference standard solutions, thus finishing up with two results each.

If the solutions of the reference standard only are exchanged between the two assistants and these are now plated out against the proposed new standard solutions, this will generate a further four results, making a total of eight results for the first day. This same procedure repeated on another day

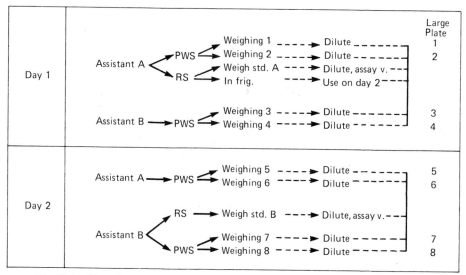

Fig. 12.1. Flow diagram showing the calibration sequence of a proposed working standard (PWS) when only a single reference standard (RS) is available. std., Standard; v, versus.

will again yield 16 results altogether. Another flow diagram giving a summary of this system is found in Figure 12.2.

The next step is to evaluate all the individual results and combine them into a suitable mean potency. The simplest approach would be to work out a straightforward arithmetic average with standard deviation and confidence limits. Other statistical calculations are also possible, such as a test for heterogeneity by the χ^2 test or the F test. Examples of these calculations are given by Hewitt (1977).

Express the mean potency obtained to three significant figures, and label each filled and properly sealed container with the name, the potency and the batch number of the standard, the date of preparation, the date of expiry, and the storage requirement. The assay working standard should normally be stored in the refrigerator or at a lower temperature, such as $-20°C$.

In addition to potency determination by bioassay, the working standard should be as fully investigated as possible by both chemical and physical means as well. Supporting information to show purity, impurities, moisture content, stability, and the like, should be lodged with the keeper of the assay working standard.

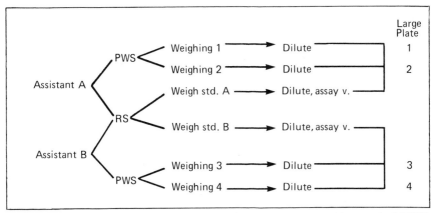

Fig. 12.2. Flow diagram showing the calibration of a proposed working standard (PWS) when two reference standards (RS) are available. Daily plan. std., Standard; v, versus.

12.6 Replication for Collaborative Assays

Setting up national or international reference standards calls for even greater replication than that suggested for internal working standards. As many expert laboratories as possible should be asked to participate. Collaborating laboratories should be encouraged to use alternative methods of assay apart from the one recommended by the central organizing laboratory. Indeed, each participant should concentrate on the assay method in which the laboratory has the greatest experience; for example, if the laboratory has been set up to use the turbidimetric method of assay, then this should be the method of choice. Three or more dose levels are usually recommended for this type of assay, but this is not essential as the potency information provided by the main three-dose level assay stream can often be supplemented by two-dose level assays, especially if the assay staff have not handled three-dose level assays before.

These assays are normally organized by government control laboratories or by large pharmaceutical companies, and full instructions to carry out large-scale collaborative assays are outside the intended scope of this book. The reader is referred to Miles (1953) and Lightbown *et al.* (1974).

References

Coomber, P. A., Jefferies, J. P., and Woodford, J. D. (1982). *Analyst* **107**, 1451.
Hewitt, W. (1977). "Microbiological Assay." Academic Press, New York.

Lees, K. A., and Tootill, J. P. R. (1955). *Analyst* **80,** 95.

Lightbown, J. W., Thomas, A. H., Grab, B., and Dixon, H. (1974). "The Second International Reference Preparation of Neomycin." Medical Research Council, London.

Miles, A. A. (1953). *J. Am. Pharm. Assoc.* (Sci. Ed.) **42,** 226.

Simpson, J. S. (1963). *In* "Analytical Microbiology" (F. W. Kavanagh, ed.) Vol. I. Academic Press, New York.

DISINFECTANTS AND DETERGENTS SUITABLE FOR USE IN MICROBIOLOGICAL ASSAY LABORATORIES

A1.1 Disinfectants

General Guidelines

(1) Disinfectants cannot be relied on to replace sterilization processes and may be used only as indicated in the text.

(2) Diluted aqueous solutions must be prepared fresh as required and must not be stored any longer than necessary.

(3) Examples of disinfectants given below may not be available worldwide, in which case locally available disinfectants may be used after thorough testing.

(4) Before a new disinfectant can be put into use, it must be properly tested by either the method of Kelsey and Maurer (1966), the rapid evaluation method of Mossel (1963), or the AOAC (1984) use-dilution test.

(5) Obtain as much information as possible from the manufacturer but do not accept their recommendations without checking their claims in your own laboratory.

Savlon hospital concentrate: Active ingredients are chlorhexidine gluconate solution BP 7.5% (v/v) [1.5% (w/v)] and cetrimide BP 15% (w/v). Manufactured by Imperial Chemical Industries, Alderley Park, Macclesfield, SK10 4TF England.

Hycolin: A broad-spectrum disinfectant. Active ingredients are synthetic phenolic derivatives. Manufactured by William Pearson Ltd., Clough Road, Hull HU6 7QA.

Stericol concentrate: Dilute to 1% (v/v) for most applications. Manufactured by Sterling Industrial, Chapeltown, Sheffield S30 4YP, England.

Duet: A granular detergent sanitizer for hand cleaning, working surfaces, and utensils. Made by Diversey UK Ltd.

Cidex: An activated glutaraldehyde solution (NFXVI/USPXXI) Manufactured by Surgikos Ltd., Kirkton Campus, Livingston, Scotland.

A1.2 Detergents

Pyroneg: For general use prepare a 0.3% (w/v) solution in water. Made by Diversey UK Ltd.

Decon 90: Phosphate-free, totally rinsable below monomolecular-layer level, and biodegradable. Made by Decon Laboratories Ltd., Conway Street, Hove, Sussex BN3 2ZZ, England.

Neodisher: Made by Medicell International Ltd., 239 Liverpool Road, London N1 1LX, England.

Rapidex: Made by Rapidex, 16 Jacobswell Mews, George Street, London W1H 6BD, England.

Lab-Brite: Made by British Hydrological Corporation.

Micro: Made by International Products Corporation, 1 Church Row, Chislehurst, Kent BR7 5PG, England (branch of International Products Corporation, P. O. Box 118, Trenton, New Jersey 08601–0118).

References

AOAC, Williams, S. ed. (1984). "Official Methods of Analysis of the Association of Official Analytical Chemists" 14th. Ed., p. 67, AOAC Inc., Arlington, Virginia 22209.

Kelsey, J. C., and Maurer, Isobel M. (1966). *Bull. Min Health and PHLS*, **25**, 180.

Mossel, D. A. A. (1963). *Lab. Practice* (London) **12**, 898.

APPENDIX 2

SUPPLIERS OF BACTERIOLOGICAL CULTURE MEDIA

Difco Laboratories. Head office: P.O. Box 1058, Detroit, Michigan 48232. P.O. Box 14B, Central Avenue, East Molesey, Surrey KT8 0SE England.

Oxoid. Head office: Oxoid Ltd., Wade Road, Basingstoke, Hampshire, RG24 0PW, England. Oxoid USA Inc., 9017 Red Branch Road, Columbia, Maryland 21045.

BBL. Head office: Division of Becton Dickinson and Company, Cockeysville, Maryland 21030. Becton Dickinson UK Ltd., Between Towns Road, Cowley, Oxford OX4 3LY, England.

Gibco Ltd. P.O. Box 35, Trident House, Renfrew Road, Paisley, Renfrewshire PA3 4EF, Scotland, Branches in Belgium, France, Netherland, Germany, and Switzerland.

Pfizer Diagnostics. Chas. Pfizer & Co., Inc., 300 West Street, New York 10036.

Merck. Darmstadt, Germany.

Lab M. (London Analytical & Bacteriological Media Ltd.) Lab M Ltd., Ford Lane, Salford, Lancashire M6 6PB, England.

Mast Diagnostics Ltd. Mast House, Derby Road, Bootle, Merseyside L20 1EA, England.

FORMULAS AND PREPARATION OF BACTERIOLOGICAL CULTURE MEDIA

The majority of culture media described here can be obtained in a dehydrated form from most of the media manufacturers listed in Appendix 2.

In addition to the media given in this Appendix, assay cultures are often maintained on slants of the assay medium as described under the appropriate assay sections. Ingredients shown in the following formulas are per liter of medium.

Blood Agar Base

Proteose peptone	15.0 g
Papain digest of liver	2.5 g
Yeast extract	5.0 g
Sodium chloride	5.0 g
Bacteriological agar	12.0 g
Approximate final pH 7.4	

Suspend the ingredients in cold distilled water for about 30 minutes, stirring frequently. Heat to boiling to dissolve, fill about 15-ml amounts into 20-ml universal bottles fitted with screw caps or into suitable test tubes, and sterilize at 121°C for 15 minutes. Allow the agar to set in a sloping position after autoclaving.

Microassay Culture Agar

Yeast extract	20.0 g
Proteose peptone No. 3	5.0 g
Dextrose	10.0 g
Potassium dihydrogen phosphate	2.0 g
Sorbitan monooleate complex	0.1 g
Bacteriological agar	10.0 g
Approximate final pH 6.7	

Suspend the ingredients in cold distilled water for about 30 minutes stirring frequently. Heat to boiling to dissolve completely. Distribute in 10-ml amounts into 20-ml universal bottles fitted with screw caps or into suitable test tubes, and sterilize at 121°C for 15 minutes. After autoclaving, allow the agar to set in an upright position.

Microinoculum Broth

Yeast extract	20.0 g
Proteose peptone No. 3	5.0 g
Dextrose	10.0 g
Potassium dihydrogen phosphate	2.0 g
Sorbitan monooleate complex	0.1 g
Approximate final pH 6.7	

Dissolve the ingredients in distilled water, applying slight heat if necessary. Fill in 10-ml amounts into 20-ml universal bottles fitted with screw caps or into suitable test tubes, and sterilize at 121°C for 15 minutes.

Nutrient Agar

Beef extract	3.0 g
Bacteriological peptone	5.0 g
Bacteriological agar	15.0 g
Approximate final pH 6.8	

Suspend the ingredients in cold distilled water for about 30 minutes, stirring frequently. Heat to boiling to dissolve, fill in about 15-ml amounts into 20-ml universal bottles fitted with screw caps or into suitable test tubes, and sterilize at 121°C for 15 minutes. After autoclaving, allow to set in an upright position. Melt down as required to pour the medium into petri dishes.

0.1% Peptone Water

Dissolve 1.0 g of bacteriological peptone and 0.89 g of sodium chloride in 1 liter of distilled water, and adjust to about pH 7.1. Dispense into suitable bottles in 100-ml amounts, and sterilize at 121°C for 15 minutes.

Sabouraud Dextrose Agar

Mycological peptone	10.0 g
Dextrose	40.0 g
Bacteriological agar	15.0 g
Approximate final pH 5.6	

Suspend the ingredients in cold distilled water for about 30 minutes, stirring frequently. Heat to boiling to dissolve, fill in about 15-ml amounts into 20-ml universal bottles fitted with screw caps or into suitable test tubes, and sterilize at 121°C for 15 minutes. After autoclaving, allow to set in a sloping position for slants or in an upright position for melting down later to pour into petri dishes.

Saline TS USP

Dissolve 9.0 g of sodium chloride in distilled water to make 1000 ml. Fill in convenient amounts and sterilize.

Soybean–Casein Digest Medium USP

Pancreatic digest of casein	17.0 g
Papaic digest of soybean meal	3.0 g
Sodium chloride	3.0 g
Dipotassium phosphate	2.5 g
Dextrose	2.5 g
Approximate final pH 7.3	

Dissolve the solids in distilled water, applying slight heat to effect solution. Fill in 10- to 15-ml amounts into 20-ml universal bottles fitted with screw caps or into suitable test tubes, and sterilize at 121°C for 15 minutes.

Soybean–Casein Digest Agar Medium

Pancreatic digest of casein	15.0 g
Papaic digest of soybean meal	5.0 g
Sodium chloride	5.0 g
Bacteriological agar	15.0 g
Approximate final pH 7.3	

Suspend the ingredients in cold distilled water for about 30 minutes, stirring frequently. Heat to boiling to dissolve, fill in about 15-ml amounts into 20-ml universal bottles fitted with screw caps or into suitable test tubes, and sterilize at 121°C for 15 minutes. After autoclaving, allow to set in a sloping position for slants or in an upright position for melting down later to pour into petri dishes.

Sporulating Medium

Nutrient broth No. 2 (Oxoid)	3.125 g
Manganese chloride ($MnCl_2 \cdot 4H_2O$)	0.030 g
Dipotassium phosphate	0.250 g
Bacteriological agar	12.0 g
Approximate final pH 7.0	

Suspend the ingredients in cold distilled water for about 30 minutes, stirring frequently. Heat to boiling to dissolve, dispense in 130- to 140-ml amounts into 500-ml medical flat bottles, and sterilize at 121°C for 15 minutes. After autoclaving, lay flat on one side and allow to solidify.

LIST OF MEDIA AND DILUENTS USED IN ANTIBIOTIC ASSAYS

The media and diluents used in the assay of antibiotics in Chapters 9 and 11 correspond to the numbering in the Code of Federal Regulations (CFR 21, 1985, paragraph 436.102). Media not described in CFR have been numbered in sequence starting with medium 40.

A4.1 Media

Instead of preparing the media from individual ingredients, it is often more economical and more convenient to make them from dehydrated mixtures which, when reconstituted with distilled water according to the manufacturers' instructions, have the same composition as the media here listed. Manufacturers of assay media can be found in Appendix 2.

Note that media 1 and 2 can have the final pH adjusted to various values depending on the particular assay. For example, medium 1 occurs with pH values of 6.5, 7.5, and 7.9. Medium 11 has the same formula as medium 1 but has a final pH of 7.9. Similarly, medium 2 occurs with pH 5.9 (medium 8), 6.5, 7.5, 7.9 (medium 5), and 8.5.

Medium 1

Bacteriological peptone	6.0 g
Pancreatic digest of casein	4.0 g
Yeast extract	3.0 g
Beef extract	1.5 g
Dextrose	1.0 g
Agar	15.0 g
Distilled water to	1000.0 ml
pH 6.5 – 6.6 after sterilization	

Medium 2

Bacteriological peptone	6.0 g
Yeast extract	3.0 g
Beef extract	1.5 g

Agar	15.0 g
Distilled water to	1000.0 ml
pH 6.5 – 6.6 after sterilization	

Medium 3

Bacteriological peptone	5.0 g
Yeast extract	1.5 g
Beef extract	1.5 g
Sodium chloride	3.5 g
Dextrose	1.0 g
Dipotassium phosphate	3.68 g
Potassium dihydrogen phosphate	1.32 g
Distilled water to	1000.0 ml
pH 6.95 – 7.05 after sterilization	

Medium 5

Medium 5 is the same as medium 2, except adjust the final pH to 7.8 – 8.0 after sterilization.

Medium 8

Medium 8 is the same as medium 2, except adjust the final pH to 5.8 – 6.0 after sterilization.

Medium 9

Pancreatic digest of casein	17.0 g
Papaic digest of soybean	3.0 g
Sodium chloride	5.0 g
Dipotassium phosphate	2.5 g
Dextrose	2.5 g
Agar	20.0 g
Distilled water to	1000.0 ml
pH 7.2 – 7.3 after sterilization	

Medium 10

Medium 10 is the same as medium 9, except:

Agar	12.0 g
Polysorbate 80	10.0 ml
pH 7.2 – 7.3 after sterilization	

Add polysorbate 80 after boiling the medium to dissolve the agar.

Medium 11

Medium 11 is the same as medium 1, except adjust the final pH to 7.8–8.0 after sterilization.

Medium 13

Bacteriological peptone	10.0 g
Dextrose	20.0 g
Distilled water to	1000.0 ml
pH 5.6–5.7 after sterilization	

Medium 19

Bacteriological peptone	9.4 g
Yeast extract	4.7 g
Beef extract	2.4 g
Sodium chloride	10.0 g
Dextrose	10.0 g
Agar	23.5 g
Distilled water to	1000.0 ml
pH 6.0–6.2 after sterilization	

Medium 32

Prepare as medium 1, except add 300 mg of hydrated manganese sulfate ($MnSO_4 \cdot H_2O$) to each liter of medium.

Medium 34

Glycerol	10.0 g
Bacteriological peptone	10.0 g
Beef extract	10.0 g
Sodium chloride	3.0 g
Distilled water to	1000.0 ml
pH 7.0 after sterilization	

Medium 35

Prepare as medium 34, except add 17.0 g of agar to each liter of medium.

Medium 36

Pancreatic digest of casein	15.0 g
Papaic digest of soybean	5.0 g
Sodium chloride	5.0 g
Agar	15.0 g
Distilled water to	1000.0 ml
pH 7.3 after sterilization	

Medium 37

Pancreatic digest of casein	17.0 g
Soybean peptone	3.0 g
Dextrose	2.5 g
Sodium chloride	5.0 g
Dipotassium phosphate	2.5 g
Distilled water to	1000.0 ml

pH 7.3 after sterilization

Medium 38

Bacteriological peptone	15.0 g
Papaic digest of soybean meal	5.0 g
Sodium chloride	4.0 g
Sodium sulfite	0.2 g
L-cystine	0.7 g
Dextrose	5.5 g
Agar	15.0 g
Distilled water to	1000.0 ml

pH 7.5 after sterilization

Medium 40

Bacteriological peptone	5.0 g
Beef extract	2.4 g
Agar	15.0 g
Distilled water to	1000.0 ml

pH 6.5 after sterilization

Medium 41

Prepare as medium 2, except add 5.0 g of sodium chloride to each liter of medium.

Medium 42

Bacteriological peptone	5.0 g
Beef extract	2.4 g
Trisodium citrate	10.0 g
Sodium chloride	5.0 g
Agar	15.0 g
Distilled water to	1000.0 ml

pH 6.8 after sterilization

Medium 43

Bacteriological peptone	5.0 g
Beef extract	2.4 g
Trisodium citrate	10.0 g
Agar	15.0 g
Distilled water to	1000.0 ml

pH about 7.0 after sterilization

Note. Attention is drawn to the different pH values that media 40–43 must be adjusted to just before adding the seeding inoculum. The various conditions are summarized in Table A4.1.

Table A4.1

pH Adjustment Necessary for Media 40–43 to Make Them Acceptable for the Assays as Indicated

Medium	pH	Section	Assay
40	6.7	9.23	Oxytetracycline
	7.2	9.3	Amoxicillin
	7.5	9.18	Dihydrostreptomycin
	7.5	9.27	Streptomycin
41	6.5	9.17	Cloxacillin
	6.5	9.22	Nystatin
	7.0	9.8	Carbenicillin
42	6.5	9.9	Cefazolin
	6.5	9.12	Cefuroxime
	7.0	9.10	Cefotaxime
	7.2	9.13	Cephalexin
	7.5	9.24.2	Penicillin G, method 2
	7.9	9.26	Phenylmercuric acetate
	7.9	9.29	Thiomersal
43	6.0	9.11	Ceftazidime
	6.0	9.14	Cephaloridine
	7.0	9.24.3	Penicillin G, method 3

A4.2   Diluents

Solution 1   (1% potassium phosphate buffer, pH 6.0)

Dipotassium phosphate	2.0 g
Potassium dihydrogen phosphate	8.0 g
Distilled water to	1000.0 ml

Adjust with 18 *N* phosphoric acid or 10 *N* potassium hydroxide to yield a pH of 5.95 – 6.05 after sterilization.

Solution 3   (0.1 *M* potassium phosphate buffer, pH 8.0)

Dipotassium phosphate	2.0 g
Potassium dihydrogen phosphate	8.0 g
Distilled water to	1000.0 ml

Adjust with 18 *N* phosphoric acid or 10 *N* potassium hydroxide to yield a pH of 7.9 – 8.1 after sterilization.

Solution 4   (0.1 *M* potassium phosphate buffer, pH 4.5)

Potassium dihydrogen phosphate	13.6 g
Distilled water to	1000.0 ml

Adjust with 18 *N* phosphoric acid or 10 *N* potassium hydroxide to yield a pH of 4.45 – 4.55 after sterilization.

Solution 6   (10% potassium phosphate buffer, pH 6.0)

Dipotassium phosphate	20.0 g
Potassium dihydrogen phosphate	80.0 g
Distilled water to	1000.0 ml

Adjust with 18 *N* phosphoric acid or 10 *N* potassium hydroxide to yield a pH of 10.4 – 10.6 after sterilization.

Solution 10   (0.2 *M* potassium phosphate buffer, pH 10.5)

Dipotassium phosphate	35.0 g
10 *N* potassium hydroxide	2.0 ml
Distilled water to	1000.0 ml

Adjust with 18 *N* phosphoric acid or 10 *N* potassium hydroxide to yield a pH of 10.4 – 10.6 after sterilization.

Solution 13   (0.01 *N* methanolic hydrochloric acid)

1.0 *N* hydrochloric acid	10.0 ml
Methanol to	1000.0 ml

Solution 15   (80% 2-propanol solution)

2-Propanol	800.0 ml
Distilled water to	1000.0 ml

Solution 16   (0.1 *M* potassium phosphate buffer, pH 7.0)

Dipotassium phosphate	13.6 g
Potassium dihydrogen phosphate	4.0 g
Distilled water to	1000.0 ml

Adjust with 18 *N* phosphoric acid or 10 *N* potassium hydroxide to yield a pH of 6.8 – 7.2 after sterilization.

Solution 20   (0.07 *M* sodium phosphate buffer, pH 7.5)

Disodium phosphate	9.78 g
Potassium dihydrogen phosphate	1.85 g
Distilled water to	1000.0 ml

Adjust with 18 *N* phosphoric acid or 10 *N* potassium hydroxide to yield a pH of 7.4 – 7.6 after sterilization.

Solution 21   (0.1% peptone diluent)

| Bacteriological peptone | 20.0 g |
| Distilled water to | 1000.0 ml |

Distribute in convenient amounts and sterilize at 121 °C for 15 minutes.

Solution 22   (2% hydrochloric acid in methanol)

| Concentrated hydrochloric acid | 20.0 ml |
| Methanol to | 1000.0 ml |

WARNING: Use safety pipette and eye protection when handling concentrated acid.

APPENDIX 5

SOURCES OF REFERENCE SUBSTANCES AND CULTURES

A5.1 Reference Substances

Some sources from which materials may be obtained for use as reference standards are given below. These range from the source for the ultimate International Biological Standards to sources of genuine substances which may be used as working standards after calibration against a master standard.

A5.1.1 International Biological Standards and International Biological Reference Preparations

About 40 of these antibiotic substances, which are described in Section 3.2 of Chapter 3, are held and distributed for the World Health Organization by the International Laboratory for Biological Standards at the National Institute for Biological Standardization and Control (Blanche Lane, South Mimms, Potters Bar EN6 3QG, England). The range of substances includes those antibiotics that have an established place in medical treatment and for which biological standardization is the generally accepted procedure. The range does not include the natural and semisynthetic penicillins or the cephalosporins.

The substances are available only in very limited quantities and are intended as master standards against which national standards may be calibrated. They are available free of charge to laboratories of national regulatory authorities.

A5.1.2 British Biological Standards and Reference Preparations

These are generally identical with the corresponding international preparations. Additionally there is a small number of antibiotic substances for which no corresponding international reference substance has been established. They are distributed from the same address as the international standards and preparations.

A5.1.3 European Pharmacopoeia Commission Reference Substances

These include a number of antibiotics, some of which are intended as chemical reference substances and others as biological reference substances.

They are distributed by the Secretariat of the European Pharmacopoeia Commission, Council of Europe, 67000, Strasbourg, France.

A5.1.4   United States Pharmacopoeia Reference Substances

This extensive list of pharmaceutical substances includes about 30 antibiotics and 10 vitamins. Some of these are intended as chemical reference substances and others as biological reference substances. In contrast to the International and British lists, this does include substances intended for the biological standardization of the penicillins and cephalosporins.

These substances are distributed by USP Reference Standards, 4630 Montgomery Avenue, Bethesda, Maryland 20014.

A5.1.5   International Chemical Reference Substances

These pharmaceutical substances are intended primarily for use in infrared identification, in chromatographic tests and assays, and in spectrophotometric assay methods. However, the International Pharmacopoeia prescribes the use of a few of these substances in the biological assay of pharmaceutical dosage forms of some of the semisynthetic antibiotics (but not for the assay of the pharmaceutical substances themselves, which are more appropriately assayed by chemical/physicochemical means).

These substances are distributed by the World Health Organization Collaborating Centre for Chemical Reference Substances, Apoteksbolaget ab centrallaboratoriet S-105 14 Stockholm, Sweden.

A5.1.6   Other Sources of Materials for Use as Working Standards

An "in-house" working standard may be established by setting aside for this purpose a quantity of a genuine sample of the material for which a standard is required and calibrating it against one of the recognized reference substances. Sometimes this genuine sample may be obtained from a primary manufacturer of the material. Alternatively, it may be obtained from a supplier of chemical reagents. Two such sources in Britain are British Drug Houses, Poole, Dorset and Sigma Chemicals, Poole, Dorset.

A5.2   Cultures

The cultures mentioned in this book may be obtained from the following sources:

American Type Culture Collection (ATCC), 12301 Parklawn Drive, Rockville, Maryland. 20852

Commonwealth Mycological Institute (IMI), Ferry Lane, Kew TW9 3AF, England

National Collection of Industrial Bacteria (NCIB), Torry Research Station, P.O. Box 31, 135 Abbey Road, Aberdeen AB9 8DG, Scotland

National Collection of Type Cultures (NCTC), Central Public Health Laboratory, Colindale Avenue, London NW9 5HT, England

National Collection of Yeast Cultures (NCYC), The Brewing Research Foundation, Lyttel Hall, Nutfield, Redhill RH1 4HY, England

APPENDIX 6

EXPRESSIONS FOR THE CALCULATION OF E AND F FOR POTENCY ESTIMATION FROM PARALLEL LINE ASSAYS

For two dose levels:

$$E = \tfrac{1}{2}[(S_2 + T_2) - (S_1 + T_1)]$$

$$F = \tfrac{1}{2}[(T_2 + T_1) - (S_2 + S_1)]$$

For three dose levels:

$$E = \tfrac{1}{4}[(S_3 + T_3) - (S_1 + T_1)]$$

$$F = \tfrac{1}{3}[(T_3 + T_2 + T_1) - (S_3 + S_2 + S_1)]$$

For four dose levels:

$$E = \tfrac{1}{20}[3(S_4 + T_4) + (S_3 + T_3) - (S_2 + T_2) - 3(S_1 + T_1)]$$

$$F = \tfrac{1}{4}[(T_4 + T_3 + T_2 + T_1) - (S_4 + S_3 + S_2 + S_1)]$$

For five dose levels:

$$E = \tfrac{1}{20}[2(S_5 + T_5) + (S_4 + T_4) - (S_2 + T_2) - 2(S_1 + T_1)]$$

$$F = \tfrac{1}{5}[(T_5 + T_4 + T_3 + T_2 + T_1) - (S_5 + S_4 + S_3 + S_2 + S_1)]$$

For six dose levels:

$$E = \tfrac{1}{70}[5(S_6 + T_6) + 3(S_5 + T_5) + (S_4 + T_4) - (S_3 + T_3) - 3(S_2 + T_2) - 5(S_1 + T_1)]$$

$$F = \tfrac{1}{6}[(T_6 + T_5 + T_4 + T_3 + T_2 + T_1) - (S_6 + S_5 + S_4 + S_3 + S_2 + S_1)]$$

BASIC BACTERIOLOGICAL STAINING TECHNIQUES

A7.1 Introduction

The brief guidelines given here are intended as a reminder to aid in the identification of assay organisms. For fuller details of bacteriological staining techniques, consult standard textbooks of bacteriology.

A7.2 Preparation and Fixation of Bacterial Smears

Use grease-free, ultraclean bacteriological slides. Transfer a loopful of a freshly grown broth culture (24 – 48 hours) to the middle of a bacteriological slide and spread the liquid evenly over a large area. To examine a culture grown on a solid agar surface, first transfer a loopful of sterile distilled water to the middle of a slide, then touch the surface of a single colony with the edge of a sterile bacteriological loop and emulsify the just-visible trace of bacterial growth on the loop in the drop of water, spreading evenly over a large area of the slide. Allow the smear to dry completely at room temperature, and fix the bacterial film by passing the slide two or three times through a Bunsen flame so that the slide feels hot to the touch. Use tongs for holding the slide to avoid burning the fingers. Cultures grown in broth containing glucose generally need more heat for fixing, but beware of burning the smear. Alternately, a thermal block may be used instead of Bunsen flame.

A7.2.1 Gram Staining (Hucker's modification: Hucker and Conn, 1927)

Cover the fixed smear with crystal violet solution and allow to stand for 1 minute.

Rinse the slide carefully with gently running tap water. Avoid splashing.

Cover the slide with Lugol's iodine solution and allow to stand for 1 minute. Rinse gently with tap water.

Decolorize with 95% ethanol by flooding the slide two or three times or until no more color is removed. Rinse with tap water. *CAUTION:* Extinguish naked flames.

Counterstain with safranine solution for 1 minute, rinse with tap water, and dry the slide with a clean blotting paper (filter paper).

Examine the smear under the oil immersion lens of a microscope.

Gram-positive organisms are stained deep purple (e.g., *Bacillus, Staphylococcus, Micrococcus*) and gram-negative organisms appear light pink (e.g., *Escherichia*).

A7.2.2   Spore Staining   (Schaeffer and Fulton, 1933)

Flood the fixed smear with malachite green (5% aqueous solution), and heat the underside of the slide until steam rises. Repeat three or four times within 30 seconds, but do not boil. (A cotton wool ball wrapped around a bacteriological loop, saturated with ethanol and set on fire, is generally used for heating the slide. *CAUTION:* Do not leave the container of ethanol near the naked flame.)

Wash off excess stain under gently running tap water for about 30 seconds.

Apply safranine [0.5% (w/v) aqueous] solution for 30 seconds.

Rinse with water and blot dry.

Examine under the oil immersion lens of the microscope.

The bacterial spores should be stained green, sporangia and vegetative cells pink.

A7.3   Preparation of Solutions for Staining

A7.3.1   Gram Stain

Crystal Violet – Oxalate Solution
   Solution A

Crystal violet (90% dye content)	2.0 g
Ethanol [95% (v/v)]	20.0 ml

   Solution B

Ammonium oxalate	0.8 g
Distilled water	80.0 ml

Mix solutions A and B in equal volumes.

CAUTION: Extinguish all naked flames before using ethanol.

Lugol's Iodine Solution

Iodine	1.0 g
Potassium iodide	2.0 g
Distilled water	300.0 ml

Counterstain

| Safranine solution [2.5% (w/v) in 95% ethanol] | 10 ml |
| Distilled water | 100 ml |

A7.3.2 Spore Stain

Malachite Green Solution

| Malachite green | 5.0 g |
| Distilled water to | 100.0 ml |

Allow to stand for 30 minutes and filter.

Safranine Solution

| Safranine | 0.5 g |
| Distilled water to | 100.0 ml |

References

Hucker, G. J., and Conn, H. J. (1927). *N.Y. State Agric. Exp. Stn. Tech. Bull.* p. 128.
Schaeffer, A. B., and Fulton, B. B. (1933). *Science* 77, 194.

APPENDIX 8

SOME PRACTICAL ASSAY DESIGNS

A8.1 4 × 4 Latin Squares

Suitable for one sample and one standard at two dose levels each. Four replications for each solution.

Design 1	Design 2	Design 3	Design 4
3 1 4 2	4 1 3 2	1 3 4 2	2 3 1 4
2 4 3 1	2 4 1 3	4 2 1 3	1 4 2 3
1 3 2 4	3 2 4 1	3 4 2 1	3 2 4 1
4 2 1 3	1 3 2 4	2 1 3 4	4 1 3 2

Design 5	Design 6	Design 7	Design 8
1 3 2 4	2 4 3 1	3 2 4 1	4 2 1 3
3 1 4 2	4 2 1 3	1 3 2 4	3 4 2 1
2 4 3 1	3 1 4 2	4 1 3 2	2 1 3 4
4 2 1 3	1 3 2 4	2 4 1 3	1 3 4 2

Design 9	Design 10	Design 11	Design 12
4 3 1 2	2 4 1 3	4 2 3 1	3 1 2 4
1 2 4 3	1 3 4 2	1 4 2 3	1 2 4 3
3 1 2 4	4 2 3 1	2 3 1 4	4 3 1 2
2 4 3 1	3 1 2 4	3 1 4 2	2 4 3 1

Design 13	Design 14	Design 15	Design 16
2 1 4 3	1 3 2 4	2 1 3 4	3 4 2 1
4 3 2 1	4 2 1 3	3 4 2 1	1 3 4 2
3 4 1 2	3 1 4 2	4 2 1 3	2 1 3 4
1 2 3 4	2 4 3 1	1 3 4 2	4 2 1 3

Design 17	Design 18	Design 19	Design 20
1 2 4 3	4 3 1 2	3 4 1 2	2 1 3 4
4 3 1 2	3 1 2 4	2 1 4 3	1 3 4 2
2 4 3 1	2 4 3 1	1 3 2 4	4 2 1 3
3 1 2 4	1 2 4 3	4 2 3 1	3 4 2 1

A8.2 6 × 6 Latin Squares

Suitable for two samples and one standard at two dose levels each or one sample and one standard at three dose levels each. Six replications for each solution.

Design 1	Design 2	Design 3
2 5 6 3 4 1	2 4 6 1 5 3	6 1 4 3 2 5
4 6 2 1 5 3	5 1 3 4 2 6	3 6 1 5 4 2
1 3 5 4 2 6	3 5 2 6 1 4	1 3 5 2 6 4
3 1 4 5 6 2	4 6 1 5 3 2	4 5 2 1 3 6
5 2 1 6 3 4	1 3 4 2 6 5	2 4 3 6 5 1
6 4 3 2 1 5	6 2 5 3 4 1	5 2 6 4 1 3

Design 4	Design 5	Design 6
3 5 2 4 1 6	4 1 6 3 5 2	5 4 6 3 1 2
1 4 6 5 2 3	2 6 3 4 1 5	2 6 5 1 3 4
6 3 5 2 4 1	5 3 2 1 4 6	4 2 1 6 5 3
4 2 1 3 6 5	6 5 1 2 3 4	3 1 2 5 4 6
2 1 3 6 5 4	3 2 4 5 6 1	1 3 4 2 6 5
5 6 4 1 3 2	1 4 5 6 2 3	6 5 3 4 2 1

Design 7	Design 8	Design 9
4 2 1 3 6 5	2 6 3 4 1 5	1 4 2 5 3 6
6 1 2 5 4 3	5 3 2 1 4 6	3 6 1 2 5 4
1 5 4 2 3 6	6 5 1 2 3 4	6 5 4 3 1 2
2 6 3 1 5 4	3 2 4 5 6 1	5 3 6 4 2 1
5 3 6 4 2 1	1 4 5 6 2 3	2 1 3 6 4 5
3 4 5 6 1 2	4 1 6 3 5 2	4 2 5 1 6 3

Design 10	Design 11	Design 12
1 4 5 6 3 2	3 5 6 4 2 1	4 6 1 5 3 2
4 5 6 3 2 1	2 3 5 1 6 4	3 5 6 4 2 1
5 1 4 2 6 3	6 1 3 2 4 5	2 3 5 1 6 4
2 6 3 1 5 4	5 4 2 3 1 6	6 1 3 2 4 5
6 3 2 4 1 5	1 2 4 6 5 3	5 4 2 3 1 6
3 2 1 5 4 6	4 6 1 5 3 2	1 2 4 6 5 3

Design 13	Design 14	Design 15
5 3 6 4 2 1	4 5 6 3 2 1	6 5 3 4 2 1
3 4 5 6 1 2	5 1 4 2 6 3	5 4 6 3 1 2
4 2 1 3 6 5	2 6 3 1 5 4	2 6 5 1 3 4
6 1 2 5 4 3	6 3 2 4 1 5	4 2 1 6 5 3
1 5 4 2 3 6	3 2 1 5 4 6	3 1 2 5 4 6
2 6 3 1 5 4	1 4 5 6 3 2	1 3 4 2 6 5

	Design 16							Design 17							Design 18					
1	4	5	6	2	3		3	2	1	5	4	6		2	6	5	1	3	4
4	1	6	3	5	2		1	4	5	6	3	2		4	2	1	6	5	3
2	6	3	4	1	5		4	5	6	3	2	1		3	1	2	5	4	6
5	3	2	1	4	6		5	1	4	2	6	3		1	3	4	2	6	5
6	5	1	2	3	4		2	6	3	1	5	4		6	5	3	4	2	1
3	2	4	5	6	1		6	3	2	4	1	5		5	4	6	3	1	2

	Design 19							Design 20							Design 21					
3	4	5	6	1	2		4	2	5	1	6	3		2	3	5	1	6	4
4	2	1	3	6	5		1	4	2	5	3	6		6	1	3	2	4	5
6	1	2	5	4	3		3	6	1	2	5	4		5	4	2	3	1	6
1	5	4	2	3	6		6	5	4	3	1	2		1	2	4	6	5	3
2	6	3	1	5	4		5	3	6	4	2	1		4	6	1	5	3	2
5	3	6	4	2	1		2	1	3	6	4	5		3	5	6	4	2	1

A8.3 8 × 8 Latin Squares

These are the most versatile and most economical designs. Suitable for three samples and one standard at two dose levels each, or for one sample and one standard at four dose levels each. The latter is not recommended because of the possibility of curvature of the dose–response line. Each solution is replicated 8 times. A more satisfactory arrangement is what is known as single-sample Latin square (SSLS) assays, which use two dose levels of sample and standard but each solution is replicated 16 times. For example, positions 1 and 3 are filled with the high-dose solution of the sample, while positions 2 and 4 are filled with the low-dose solution of the sample. Similarly, positions 5 and 7 are filled with standard high dose and positions 6 and 8 with standard low dose.

The same designs can also be used for three-dose level assays, known as high-medium-medium-low (HMML) assays. The high dose of the sample occupies position 1 (8 replicates), the medium dose positions 2 and 3 (16 replicates), and the low dose is in position 4 (8 replicates). Similarly, the high dose of the standard occupies position 5, the medium dose positions 6 and 7, finally the low dose is in position 8.

A selection of designs from the authors' collection follows.

	Design 1								Design 4						
3	7	6	8	2	5	4	1		1	7	4	6	8	3	2	5
8	5	3	6	4	1	7	2		6	2	8	1	7	4	5	3
4	6	2	7	1	8	5	3		8	4	6	5	1	7	3	2
1	3	4	5	8	7	2	6		3	8	1	7	5	2	4	6
2	1	8	4	5	6	3	7		5	6	2	3	4	1	8	7
7	8	5	3	6	2	1	4		7	1	3	8	2	5	6	4
6	4	1	2	7	3	8	5		4	5	7	2	3	6	1	8
5	2	7	1	3	4	6	8		2	3	5	4	6	8	7	1

Design 15

6	3	8	4	2	7	5	1
2	7	5	8	1	6	4	3
7	1	6	2	5	4	3	8
1	2	4	7	3	8	6	5
3	5	2	6	8	1	7	4
8	6	7	3	4	5	1	2
5	4	3	1	7	2	8	6
4	8	1	5	6	3	2	7

Design 22

8	3	6	5	2	7	4	1
4	1	3	8	6	2	5	7
6	8	4	1	3	5	7	2
3	6	7	2	5	4	1	8
2	5	1	3	7	6	8	4
1	7	8	6	4	3	2	5
7	2	5	4	8	1	6	3
5	4	2	7	1	8	3	6

Design 36

8	1	6	5	2	7	3	4
6	2	7	3	8	1	4	5
3	7	4	1	6	5	8	2
5	8	2	4	7	3	1	6
2	4	5	6	3	8	7	1
7	3	1	2	5	4	6	8
4	6	8	7	1	2	5	3
1	5	3	8	4	6	2	7

Design 42

5	2	6	1	7	4	3	8
6	1	5	7	3	8	4	2
8	3	2	6	4	5	1	7
4	7	1	8	6	3	2	5
1	5	4	2	8	6	7	3
7	4	8	3	2	1	5	6
3	6	7	5	1	2	8	4
2	8	3	4	5	7	6	1

Design 58

7	2	6	3	5	4	8	1
8	3	4	2	6	1	7	5
4	6	1	7	8	3	5	2
2	5	8	4	3	7	1	6
6	7	3	1	2	5	4	8
1	8	2	5	4	6	3	7
5	4	7	8	1	2	6	3
3	1	5	6	7	8	2	4

Design 63

3	7	1	8	2	6	5	4
6	8	7	5	1	4	2	3
7	3	5	2	4	8	1	6
4	6	2	3	5	1	7	8
1	2	3	4	6	7	8	5
8	5	4	7	3	2	6	1
2	4	6	1	8	5	3	7
5	1	8	6	7	3	4	2

Design 74

4	5	8	6	2	7	1	3
6	8	2	5	7	1	3	4
2	3	6	4	8	5	7	1
3	4	1	2	5	6	8	7
1	6	7	3	4	8	5	2
8	1	4	7	3	2	6	5
5	7	3	1	6	4	2	8
7	2	5	8	1	3	4	6

Design 86

2	1	3	5	8	6	7	4
3	6	4	7	5	8	2	1
4	7	2	8	6	1	5	3
7	5	1	3	4	2	6	8
6	3	5	2	1	4	8	7
8	4	7	6	2	3	1	5
1	2	8	4	7	5	3	6
5	8	6	1	3	7	4	2

Design 91

7	8	2	1	3	6	5	4
2	1	7	6	4	3	8	5
5	6	1	2	7	4	3	8
1	5	3	4	8	2	6	7
3	4	8	5	1	7	2	6
4	3	6	8	2	5	7	1
8	7	5	3	6	1	4	2
6	2	4	7	5	8	1	3

Design 96

7	3	8	1	2	6	5	4
2	4	1	6	7	8	3	5
4	5	7	2	6	3	1	8
5	2	3	7	8	1	4	6
3	8	5	4	1	2	6	7
1	7	6	8	4	5	2	3
6	1	4	5	3	7	8	2
8	6	2	3	5	4	7	1

Design 117

```
7 4 1 8 3 2 5 6
1 6 2 5 4 7 8 3
4 3 6 7 8 5 2 1
8 7 4 2 6 1 3 5
6 8 5 4 1 3 7 2
3 5 7 6 2 8 1 4
2 1 8 3 5 4 6 7
5 2 3 1 7 6 4 8
```

Design 131

```
6 4 8 1 3 5 7 2
7 1 6 4 5 8 2 3
8 2 4 5 6 7 3 1
4 8 2 7 1 3 6 5
3 6 7 8 2 1 5 4
1 3 5 2 4 6 8 7
2 5 3 6 7 4 1 8
5 7 1 3 8 2 4 6
```

Design 144

```
8 5 6 3 1 2 4 7
2 1 7 8 6 4 3 5
6 3 5 2 8 1 7 4
1 8 4 7 5 3 6 2
5 6 3 4 2 7 1 8
4 2 1 5 7 6 8 3
7 4 2 6 3 8 5 1
3 7 8 1 4 5 2 6
```

Design 151

```
5 1 4 2 8 7 6 3
2 3 6 5 1 4 8 7
8 6 5 3 4 1 7 2
4 5 7 8 3 6 2 1
7 2 1 6 5 8 3 4
6 8 3 1 7 2 4 5
1 4 8 7 2 3 5 6
3 7 2 4 6 5 1 8
```

Design 168

```
6 7 4 5 2 1 3 8
1 2 7 8 4 3 5 6
2 5 3 7 1 8 6 4
7 8 2 3 5 6 4 1
8 3 5 2 6 4 1 7
5 4 1 6 3 7 8 2
3 1 6 4 8 2 7 5
4 6 8 1 7 5 2 3
```

Design 181

```
6 7 5 1 4 3 2 8
5 2 7 4 6 8 3 1
1 6 2 3 7 4 8 5
7 3 6 5 8 2 1 4
4 8 3 6 5 1 7 2
8 1 4 2 3 7 5 6
2 4 8 7 1 5 6 3
3 5 1 8 2 6 4 7
```

Design 197

```
8 5 2 7 4 6 3 1
2 4 7 3 6 5 1 8
6 3 8 4 1 2 5 7
1 8 4 5 7 3 6 2
4 2 6 1 5 7 8 3
3 7 5 8 2 1 4 6
7 1 3 6 8 4 2 5
5 6 1 2 3 8 7 4
```

Design 198

```
3 6 5 7 4 2 8 1
5 1 2 6 8 7 3 4
8 5 6 3 1 4 7 2
6 3 1 2 7 8 4 5
7 4 8 5 2 6 1 3
4 8 7 1 3 5 2 6
1 2 4 8 5 3 6 7
2 7 3 4 6 1 5 8
```

A8.4 8 × 8 Quasi-Latin Designs

Suitable for seven samples and one standard, each at two dose levels. Each solution is replicated only four times; therefore, these are meant to be used for low-precision assays only. A selection of designs from the authors' collection follows.

Design 1

14	10	7	11	1	6	3	16
8	4	13	5	15	12	9	2
15	13	2	4	12	9	8	5
12	2	5	13	9	8	15	4
6	16	11	7	3	14	1	10
9	5	4	2	8	15	12	13
1	7	16	10	6	3	14	11
3	11	10	16	14	1	6	7

Design 2

13	15	12	2	8	3	5	10
10	12	13	5	3	2	8	15
7	1	6	16	14	9	11	4
1	7	4	14	16	11	9	6
4	6	7	11	9	16	14	1
15	13	10	8	2	5	3	12
12	10	15	3	5	8	2	13
6	4	1	9	11	14	16	7

Design 3

7	9	13	3	16	12	1	6
16	4	12	13	1	9	6	7
6	12	4	9	7	13	16	1
3	5	15	8	14	2	11	10
10	2	8	5	3	15	14	11
1	13	9	12	6	4	7	16
14	8	2	15	11	5	10	3
11	15	5	2	10	8	3	14

Design 4

16	9	13	6	7	4	12	1
6	1	7	16	13	12	4	9
13	4	6	7	16	1	9	12
7	12	16	13	6	9	1	4
2	5	11	10	3	8	14	15
11	8	10	3	2	15	5	14
10	15	3	2	11	14	8	5
3	14	2	11	10	5	15	8

Design 10

8	16	3	13	10	1	6	11
4	12	7	5	2	9	14	15
6	10	13	3	16	11	8	1
15	5	12	2	7	4	9	14
14	2	5	7	12	15	4	9
1	3	10	16	13	6	11	8
9	7	2	12	5	14	15	4
11	13	16	10	3	8	1	6

Design 11

2	15	11	14	5	7	4	10
10	11	15	4	7	5	14	2
12	9	1	8	3	13	6	16
16	1	9	6	13	3	8	12
7	4	14	15	2	10	11	5
3	8	6	1	16	12	9	13
13	6	8	9	12	16	1	3
5	14	4	11	10	2	15	7

Design 12

8	11	1	4	13	16	10	5
9	6	14	7	12	3	15	2
2	15	7	6	9	12	14	3
5	10	4	11	8	13	1	16
13	4	10	1	16	5	11	8
12	7	15	14	3	2	6	9
3	14	6	15	2	9	7	12
16	1	11	10	5	8	4	13

Design 13

9	2	12	15	5	14	3	8
6	7	13	4	16	1	10	11
4	13	7	6	10	11	16	1
11	16	10	1	13	6	7	4
8	5	3	14	12	9	2	15
15	12	2	9	3	8	5	14
14	3	5	8	2	15	12	9
1	10	16	11	7	4	13	6

Design 14

3	5	16	13	9	8	12	2
10	4	1	6	14	15	7	11
8	12	5	2	16	13	9	3
2	16	9	8	12	3	5	13
15	7	4	11	1	6	14	10
6	14	7	10	4	11	1	15
11	1	14	15	7	10	4	6
13	9	12	3	5	2	16	8

Design 17

6	15	10	1	7	12	3	14
15	12	3	6	10	1	14	7
4	13	2	9	11	8	5	16
12	1	14	15	3	6	7	10
9	4	11	8	16	13	2	5
13	8	5	4	2	9	16	11
8	9	16	13	5	4	11	2
1	6	7	12	14	15	10	3

Design 18

16	10	2	5	13	7	12	3
5	3	13	10	12	2	7	16
14	8	4	11	15	9	6	1
11	1	15	8	6	4	9	14
1	11	9	14	4	6	15	8
3	5	7	16	2	12	13	10
10	16	12	3	7	13	2	5
8	14	6	1	9	15	4	11

Design 19

14	15	8	9	6	1	11	4
9	14	15	8	1	4	6	11
12	3	2	5	10	7	13	16
3	2	5	12	13	10	16	7
5	12	3	2	7	16	10	13
8	9	14	15	4	11	1	6
2	5	12	3	16	13	7	10
15	8	9	14	11	6	4	1

Design 20

16	9	13	8	11	2	6	3
9	2	6	13	8	3	11	16
7	14	4	1	10	5	15	12
12	7	1	10	15	14	4	5
3	16	8	11	6	9	13	2
14	5	15	4	1	12	10	7
5	12	10	15	4	7	1	14
2	3	11	6	13	16	8	9

Design 39

11	14	16	7	5	4	2	9
1	8	6	13	15	10	12	3
13	12	10	1	3	6	8	15
4	5	11	16	2	7	9	14
16	9	7	4	14	11	5	2
10	15	1	6	12	13	3	8
6	3	13	10	8	1	15	12
7	2	4	11	9	16	14	5

Design 40

3	14	7	2	12	15	5	10
7	10	3	12	2	5	15	14
2	5	12	7	3	14	10	15
16	11	6	13	9	8	4	1
13	4	9	6	16	11	1	8
12	15	2	3	7	10	14	5
9	8	13	16	6	1	11	4
6	1	16	9	13	4	8	11

Design 41

16	6	3	13	10	7	1	12
5	9	8	2	11	14	4	15
1	3	12	6	7	16	10	13
10	12	13	3	16	1	7	6
4	8	15	9	14	5	11	2
7	13	6	12	1	10	16	3
11	15	2	8	5	4	14	9
14	2	9	15	4	11	5	8

Design 46

8	14	5	16	1	11	3	10
9	7	2	13	12	4	6	15
11	1	8	5	10	16	14	3
13	15	4	9	6	2	12	7
16	10	11	8	3	5	1	14
5	3	16	11	14	8	10	1
4	12	9	2	15	13	7	6
2	6	13	4	7	9	15	12

Design 47

3	2	11	7	13	6	16	10
5	14	15	9	1	4	12	8
1	4	9	15	5	14	8	12
13	6	7	11	3	2	10	16
12	9	14	4	8	15	1	5
8	15	4	14	12	9	5	1
16	7	2	6	10	11	13	3
10	11	6	2	16	7	3	13

Design 49

7	16	1	14	10	5	4	11
10	7	4	1	11	14	5	16
6	3	8	15	13	12	9	2
13	6	9	8	2	15	12	3
3	2	15	12	6	9	8	13
16	11	14	5	7	4	1	10
2	13	12	9	3	8	15	6
11	10	5	4	16	1	14	7

Design 50

10	6	13	1	3	15	12	8
5	7	4	16	2	14	9	11
15	13	12	6	10	8	1	3
2	16	7	9	11	5	4	14
11	9	16	4	14	2	7	5
3	1	6	12	8	10	13	15
8	12	1	13	15	3	6	10
14	4	9	7	5	11	16	2

A8.5 3 × 12 Rack Designs for Tube Assays

Rack design 1

34	14	33	29	4	18	9	22	16	30	19	26
32	36	6	28	5	27	12	31	25	13	1	8
20	35	3	17	2	10	15	11	7	24	23	21

Rack design 2

35	25	1	33	4	27	9	2	20	10	29	31
17	36	6	32	13	22	18	16	14	11	7	15
28	5	3	26	34	23	12	21	30	24	19	8

Rack design 3

12	10	17	25	35	16	3	36	24	1	28	11
14	15	27	21	29	6	33	2	18	31	9	23
19	26	32	13	20	4	8	34	30	22	5	7

Rack design 4

36	31	6	21	4	10	32	8	24	3	17	25
27	20	22	34	12	29	30	13	26	11	7	9
15	1	18	19	35	5	23	16	33	2	28	14

Rack design 5

9	28	5	15	30	17	20	18	10	36	26	1
7	11	19	25	22	13	32	8	24	21	29	3
14	23	12	31	35	27	16	2	6	34	33	4

Rack design 6

24	23	3	2	1	5	15	36	18	31	29	7
8	26	17	28	33	21	35	6	12	30	34	4
27	16	20	25	11	10	13	9	19	22	14	32

Rack design 7

7	26	22	6	29	32	33	8	11	1	5	30
19	14	12	34	4	17	24	2	25	31	10	36
3	13	23	18	15	27	9	16	20	28	21	35

Rack design 8

36	17	1	15	34	29	13	22	24	5	2	8
32	12	14	35	16	23	33	19	31	9	25	20
26	21	27	10	7	28	30	4	11	3	6	18

Rack design 9

4	23	21	12	27	31	16	1	8	7	9	2
18	29	33	20	26	3	32	14	25	24	34	5
36	13	22	30	15	6	35	28	11	10	19	17

Rack design 10

21	8	22	7	31	28	4	15	25	32	20	23
2	24	14	17	6	9	10	30	26	12	5	35
33	29	16	19	1	13	18	36	34	11	3	27

Rack design 11

3	16	22	27	14	19	32	12	6	20	15	21
10	36	29	24	4	7	9	2	30	5	13	18
17	28	33	25	11	26	35	1	34	31	8	23

Rack design 12

12	13	16	4	30	8	23	36	10	9	25	6
22	31	1	19	34	29	27	7	15	33	18	20
32	26	14	5	21	11	2	17	3	24	35	28

Rack design 13

16	22	9	10	6	36	14	26	2	28	11	20
5	15	8	4	24	17	1	31	13	23	21	27
34	7	25	33	30	3	29	19	32	12	35	18

Rack design 14

22	26	16	21	20	36	35	33	14	34	24	12
30	9	25	3	31	32	5	17	23	28	2	18
11	10	4	13	29	6	7	8	15	19	1	27

Rack design 15

33	11	5	28	9	17	2	8	22	36	27	20
35	7	12	15	1	18	14	4	26	10	31	16
13	21	24	25	23	34	3	32	19	6	29	30

Rack design 16

17	30	34	22	7	21	14	6	25	28	16	15
18	8	24	5	26	35	2	19	33	36	29	27
23	12	10	4	9	31	3	20	32	13	11	1

Rack design 17

23	33	36	20	32	16	30	18	7	19	14	25
13	35	4	2	3	15	11	6	26	34	29	27
10	21	1	24	17	9	22	31	8	5	28	12

Rack design 18

19	10	12	6	32	22	24	14	11	28	9	35
25	33	21	1	20	4	18	36	2	13	15	23
27	8	5	31	29	7	16	3	34	26	17	30

Rack design 19

24	8	13	15	22	17	23	34	3	21	16	32
10	20	26	31	7	12	14	6	35	2	27	36
29	33	18	25	11	28	5	1	30	19	9	4

Rack design 20

28	17	31	19	24	23	34	27	6	12	5	8
21	14	10	2	11	4	26	13	36	33	35	29
9	18	22	25	15	30	32	7	1	3	16	20

INDEX

A